F V

St. Louis Community College

Forest Park
Florissant Valley
Meramec

Instructional Resources
St. Louis, Missouri

Understanding Flying

Also by Richard L. Taylor

INSTRUMENT FLYING,
Third Revised Edition

FAIR-WEATHER FLYING

POSITIVE FLYING
(with William M. Guinther)

IFR for VFR PILOTS:
An Exercise in Survival

UNDER

STANDING FLYING

Richard L. Taylor

Illustrations by Paul C. Haynie

An Eleanor Friede Book
Macmillan Publishing Company
New York
Collier Macmillan Publishers
London

Macmillan Publishing Company
866 Third Avenue, New York, N.Y. 10022
Collier Macmillan Canada, Inc.

First Macmillan Edition 1987

Library of Congress Cataloging-in-Publication Data

Taylor, Richard L.
Understanding flying.
"An Eleanor Friede book."
Includes index.
1. Airplanes—Piloting. 2. Private flying.
I. Title.
[TL70.T363 1987]   629.132'5217   86-28488
ISBN 0-02-616660-7

Macmillan books are available at special discounts for bulk purchases for sales promotions, premiums, fund-raising, or educational use. For details, contact:
Special Sales Director
Macmillan Publishing Company
866 Third Avenue
New York, N.Y. 10022

10 9 8 7 6 5 4 3 2 1

Designed by MaryJane DiMassi

PRINTED IN THE UNITED STATES OF AMERICA

For Julie, Rich, and Mary Beth

Contents

Introduction:
Before You Begin ...

Aviation in America separates into three groups, which are easily identified: the military, the airlines, and everything else. "Everything else" is "general aviation." If present trends continue, each of these groups will grow, but each in its own way. The military is moving toward a smaller force of faster and "smarter" airplanes that carry space-age weapons. The airlines will be driven by economics to develop fleets that will be smaller and that will use less fuel to carry more passengers and cargo—the super-efficient super-jumbos. In either case, it's unlikely that tomorrow's skies will ever be blackened, or even grayed, by vast fleets of military or air-carrier airplanes.

But the general aviators . . . that's a different story. The number of light aircraft in America is growing by leaps and bounds, and, barring a national economic disaster, there's no end in sight—general aviation is on the move.

Much like the automobile in this country, the small airplane got its start at the hands of a few hardy souls who found it comfortable, if not dashing, to dress in jodhpurs, boots, leather jackets, and goggled helmets; a flowing white scarf topped off the uniform and the impression. For a long time, the pilot-superman myth prevailed, mostly because the early airplanes required a good bit more than the average man on the street could muster in the way of skill and daring, and, as likely as not, more than he was willing to put up with in the way of discomfort and inconvenience. Those old airplanes did an outstanding job of rapid transportation when compared to Model T Fords and the railroads, but they lacked creature comforts and reliability.

The old-timers will swear to the ease with which yesterday's airplanes flew, and maybe those early pilots were indeed supermen, because the designers and builders were also learning the hard way that speed and maneuverability and load-carrying capabilities often came at the expense of controllability and produced many aerodynamic surprises. Through trial and error, most of the wrinkles have been smoothed out. Today's small airplanes are fast, efficient, and very forgiving—to say nothing of the fact that the creatures involved with them can count on shirt-sleeve comfort. At least in general aviation, the pilot-superman's sun has set . . . thank you, gentlemen, and may you rest in peace.

If you're searching for the reason behind the current surge in general aviation (more people are flying, more airplanes are being built, aviation's horse-traders are doing a land-office business), look no further than the recent fuel crunches and a general improvement in the national economy. The former forced a lot of traveling people to find more efficient ways to get around, the latter provided dollars and hours for leisure-time activities, and general aviation filled both bills for many people. As a result, the base of active participation in aviation is widening, spreading to more segments of our society. Though I don't expect automobiles to be replaced by jet-powered vehicles that will take off from the back yard and fly you to Peoria at the push of a button, I am convinced that Everyman's airplane is just around one of the corners up ahead. The aura of "something special" that surrounded motor cars and drivers faded with time and numbers; I see no reason why small airplanes and pilots should not experience similar acceptance.

Given the technological improvements and simplifications that are inevitable in the face of such dynamic growth, the importance of human skills has shrunk somewhat when it comes to the safe and efficient operation of a contemporary flying machine. But until and unless we get to the button-pushing stage, flying can't be accomplished with a left foot–right foot attitude. There's going to be a need for human interpretation and skillful application of techniques in aviation for a long time to come. A basic understanding of the principles of flight and all the operations associated with flying can help a pilot handle in a safe and proper manner the countless situations that must be dealt with on even a short flight.

After twenty-two years of flying (the ten most recent of them spent in direct involvement with aviation education at a leading university), it has become clear to me that certain of those principles and operations are the important ones. I feel that, properly presented, they are the framework on which a beginning pilot can hang new pieces of knowledge and around which an experienced aviator can work to improve his aeronautical skills even more.

That is the basis on which this book has been written. I have sought to avoid discussing nonessential, dated material and situations over which the pilot has absolutely no control, and I have attempted to relate the entire discussion to practical matters wherever possible. Nearly all the examples and illustrations concerning airplanes relate to training-type aircraft; besides being the overwhelming majority of the general aviation aircraft population, the lightweight, two- and four-place airplanes are reasonably similar in their responses to pilot input and also make good laboratories in which to prove the theoretical pudding.

This book is a source of general information about airplanes and flying. There is a wealth of detailed additional material in texts and manuals, operating handbooks for specific airplanes, and the wide variety of publications available from the Federal Aviation Administration—not the least of which are the rules and regulations of aviation, for which all pilots are held responsible.

In the interest of the continued growth of aviation, I sincerely hope that after you read this book, your flying will be made more pleasant and satisfying. You should realize, however, that none of us will ever fully understand flying—only the birds have accomplished that, and, with the exception of one loquacious sea gull, they're not talking. There are too many situations to experience, too many changes taking place, for any one person to say with conviction, "I have learned all there is to know"—but by chipping away at the unknowns and learning something from each encounter, you will always be moving a little closer to understanding flying.

Part One

THE AIRPLANE
AND HOW IT FLIES

Aerodynamics
Aircraft Performance
Aircraft Loading
Propulsion Systems

Aerodynamics

I

Despite the advertising claims, there's a lot more to flying a modern airplane than just driving it off the runway and driving it back on again at the end of the trip. There's a dimensional difference that changes everything when you take to the air. That difference is the dimension of altitude. The only time an automobile climbs or descends is when the roadway climbs or descends, and since the driver has absolutely no control over it, who cares about altitude in a car? But an airplane moves vertically in addition to its left/right and fore-and-aft excursions (fore-and-aft movement is mostly fore, but changes in airspeed relate to freedom to move in that dimension). How much it moves and at what rate and in which direction is your primary concern as the pilot. When everything is steady, when there is no movement in any direction except ahead, that's straight and level, unaccelerated flight—the fastest way to get from one point to the next. When the pilot allows or causes movement to occur in any one or all three dimensions, the airplane will turn, descend, climb, speed up, slow down, or some combination of the three.

No matter the reason behind it, any movement of an airplane

is the result of *aerodynamics* ("aero" for the medium in which you're flying, and "dynamics" for the effect of the medium over the surfaces of the machine). Whether you make things happen or let them happen, aerodynamics are with you from take-off to touchdown, and frequently from chock to chock. An airplane is always subject to the forces of moving air, even when it's being driven along a taxiway.

Understanding aerodynamics as applied to modern training aircraft means understanding four areas of aircraft behavior: reaction to certain forces in flight (including takeoff and landing); the airplane's inherent stability; the purpose and effect of the flight controls (including power); and the fundamental maneuvers that, singly or in combination, produce controlled, purposeful flight.

FORCES AT WORK

Anything that moves independently through the air is under the inescapable influences of four forces—*lift, thrust, weight,* and *drag*—which together determine what the "thing" will do. An airplane flying at a constant speed, turning neither left nor right, with no gain or loss of altitude, has achieved a state of aerodynamic equilibrium: Thrust is exactly equal to drag, and lift is of the same magnitude as weight. They always act in pairs: lift and weight, thrust and drag.

Now visualize the stable-condition airplane, with the forces represented by vectors, or arrows. If the thrust vector gets longer than its opposite, drag, the airplane can do nothing but go faster. Should lift become greater than weight, the airplane will climb. When both thrust and lift increase at the same time, the airplane climbs and accelerates at the same time. You have complete control over these force vectors when you're in the pilot's seat, but if you don't understand what makes them grow or shrink, you're not really in command. One at a time, then:

The pulling-ahead force is most easily related to engine power expressed through the propeller, but is also provided by gravity in certain situations—as in Newton's headache after a red sphere deposited itself atop that scientist because gravity overpowered all the other forces (apples generate damn little lift). It is like the motive force that keeps a sailplane going— gravity again, but acting in concert with aeronautical design that allows the sailplane to glide forward many feet (sometimes as much as twenty or thirty in the very-high-performance competition models) for each foot it is pulled earthward by gravity. While he's losing altitude in that long glide, the sailplane pilot hopes he can find some air that is rising faster than he's descending; when he runs out of rising-faster air, end of flight. (Sailplaners not infrequently run out of ascending air too far away from home base to glide back, which is why all of them have trailers.)

Powered aircraft develop thrust in a number of ways, but the method of concern here is the propeller, an aerodynamic device that, due to its peculiar shape, causes air to be moved backward whenever the prop rotates. (Note that the engine doesn't provide thrust; it is the power source that turns the propeller—the thruster on training planes.) Newton's Third Law of Motion—"for every action there is an equal and opposite reaction"—now takes over, and when the air is moved backward, the equal and opposite reaction is expressed against the back of the prop blades, which are firmly attached to the crankshaft, which is part of the engine, which is bolted to the airplane, and voilà!—the whole machine moves forward. If you could stand alongside an airplane taking off or climbing, you'd see the bending of the blades (helicopter folks call it coning) as they pick up the thrust load.

Generally speaking, thrust is changed with power—but remember that the effect of thrust depends on its relation to the amount of drag present. If thrust were increased and somehow drag increased the same amount, there would be no acceleration; it's only when there's a difference between the two that the effect shows up. You could also hold thrust at some given value and increase the drag; with no actual change in thrust, but a relative change to a lower value, the airplane would slow down. Likewise, putting gravity to work by pointing the airplane downhill effectively adds to the thrust generated by the propeller, and if the resultant pull is greater than the drag, the speed increases.

DRAG

Of the four forces, drag is undoubtedly the one that's most in the hands of the aircraft designer. Even when the pilot operates the airplane as efficiently as possible, there's still the problem of pushing aside the air as it's encountered—the cross section of the machine presents some resistance to the medium. This holding-back resistance force is known as *drag*. A supersonic fighter aircraft has very little cross section, offers very little resistance to being pushed through the air at high speeds, and therefore has very little drag—but it also has very little room inside for the pilot. Ah, the great compromise of aircraft design—build a low-profile, small-cross-section airplane so it'll go fast, and your passengers develop the sardine syndrome; a roomy, fat cabin is comfortable, but it won't go very fast.

Since anything that impedes the flow of air around the airplane creates drag, the designer generally shoots for an airplane as aerodynamically "clean" as his budget will allow. When Donald Douglas thought up the DC-3, his airframe builders had no choice but to fasten her together with thousands of rivets, and the old girl wound up with a well-warted skin that helped limit her speed. (Don't run out and file the heads off the rivets—that's what holds your airplane together.) More and more airplanes—all the jets, and many of the smaller planes—have flush rivets, or bonded skins with no rivets at all. The same power with less drag means more speed.

You have no control over designed-in drag, but if you hold the airplane in a nose-high attitude, presenting more of its belly to the air, drag will increase, sometimes remarkably.

You have complete control over the drag created by lowering retractable landing wheels, extending wing flaps, or sticking your hand out the window. When you want to slow an airplane in a hurry, common sense says create all the drag possible while reducing thrust as much as you can—and it works.

Rivet heads, fat fuselages, wheels, and wing flaps produce a lot of drag without contributing anything to the production of lift; the engineers call their effect *parasite drag*. But there's another aerodynamic phenomenon, called *induced drag*. It's the price of admission to the flying business, since every airfoil creates (induces) drag in the very process of developing lift. When you arrive at a condition of flight that produces lift at the lowest cost in terms of drag, you're flying at the most efficient speed in consideration of the power available, the load being carried, the density of the air, and a number of other factors.

WEIGHT

From the moment the engine starts, an airplane gets lighter as fuel is consumed. Given a constantly decreasing weight, an airplane should become more and more efficient with each pound lost, and in general that's true—a lighter airplane will travel faster at a given power setting or travel at the same speed with less power—a happy option for the pilot.

But actual weight is only part of the story. Since weight

changes will always affect performance, real pounds are much less significant than *effective* weight—what the airplane "thinks" it weighs in flight.

When a roller coaster gets to the bottom of a hill, the cars would just as soon continue downward (Newton again—a body in motion tends to remain in motion), but the tracks have something else in mind, and you feel the effect as a very definite force pushing you down in the seat. The same thing happens as an airplane pulls out of a descent; if you were sitting on a bathroom scale at the time, you wouldn't be surprised to find that you weighed more than normal—your effective weight had increased in the pullout. This is due to centrifugal force: the tendency of anything going around in a circle to exert a pull toward the outside of the circle. (In the aerobatic maneuver known as a loop, the airplane flies in a vertical circle, but centrifugal force acts just the same. Any portion of any circle will develop some centrifugal force.) When an airplane pulls out of a dive, the entire restraining force is manifested in additional effective weight.

You'll seldom feel much effective weight in the course of normal flight training, but graduate into aerobatics and the name of the game is "pulling G's," or making the airplane perform tight turns and rapid pullouts and other exotic maneuvers that increase the centrifugal force, and therefore the apparent force of gravity. Whenever a body is at rest or experiencing no additional weight (as in level, straight flight), it is subject to the normal pull of gravity, or 1 G. Pull out of a dive rapidly enough, or cause the airplane to turn rapidly enough, or hit the bottom of the roller-coaster hill fast enough to cause the body to think its weight is twice normal, and you're pulling 2 G's. Rule of thumb: Any aircraft banked 60 degrees and held in level flight (no gain or loss of altitude) will experience a centrifugal force equal to twice the normal force of gravity—2 G's.

Some folks thoroughly enjoy the feeling of G forces; others go out of their way to avoid it. Regardless of your personal outlook, effective weight can't be ignored. If it should become greater than whatever lift is available, there's only one way the airplane can go—in the direction of the greater force (and that's usually down, and sometimes *out*).

Training aircraft are designed to withstand an effective weight (also known as *load factor*) of 3.8 times the force of gravity; those intended for more strenuous uses such as aerobatics, military training, and so on are stressed for even more G's. There's a lot more "tough" in your airplane than you'll be able to dish out in the course of pilot training, so don't be overly concerned about pulling the wings off.

There's also a force called *negative G,* which occurs whenever you find yourself on the outside of a turn. It's the same feeling you get when the roller coaster goes over the top of the hill, and it's the reason they put restraining bars in roller-coaster seats. In normal training maneuvers, a feeling of negative force usually indicates an overreaction by the pilot. Most trainers are built to withstand 1.5 G's in the negative direction, but that's to accommodate sudden downdrafts, not thrill seekers. Rapid, repeated changes from positive to negative G are guaranteed to make someone on the airplane very sick—and if that someone happens to be the pilot, there's big trouble ahead.*

LIFT

Here's what flying is all about: the queen of the quartet, the foremost of the four forces, lift. Without lift, an airplane is a fast tricycle; with it, you can break the bonds of gravity and

---

*Astronauts experience very high G forces while being shot into the sky. Once established in orbit, their loading goes to zero, at which point they are weightless—a delicate balance between gravity and centrifugal force.

realize the potential of your flying machine. Lift is everything. Almost without exception, every pressure you apply to the flight controls either increases or reduces the amount of lift. The direction in which the new force acts, or fails to act, pretty much determines what the airplane is going to do.

"Give me enough power and I'll fly a barn door!" said a jodhpured, helmet-and-goggled, scarf-trailing pilot of long ago. What he intended to use for flight controls was never well established, but his pronunciamento wasn't all that much in error. Since the production of lift depends mostly on something pushing air in one direction, causing that familiar equal-and-opposite reaction, it really doesn't matter what the something is. The lifting surfaces of modern jet fighters are little more than metal barn doors, but the engines that thrust them through the air are something else, frequently possessing more pushing power than the weight of the airplane, which means that they can "fly" straight up (no lift required, thank you).

Except for guided missiles, however, aircraft need to maintain themselves and their occupants in a state of horizontal flight, to develop lift during the transition from speeding tricycle to flying machine, and to effect a smooth, gradual change from wings to wheels during landing. The compromise has been reached in the form of the *airfoil,* an appurtenance (that's a fancy name for anything that's stuck to the airplane) that produces a lifting force when sufficient air is passed over or around it. In the case of fixed-wing aircraft, lift is produced as the wings are pushed through the air. With helicopters, the wings are rotated—same principle, different application.

The total lifting force created by an airfoil is the sum of two kinds of lift: that produced by the pushing away of air (reaction lift) and that produced by the reduction in air pressure because of the curvature of one side of the airfoil (Bernoulli's lift, in honor of the scientist who described the principle).

Reaction lift is the easier to understand. Replace wings with flat plates set at a slight angle, and you can see that when the machine is moved forward through the air, some of that air is forced downward. (Newton again—equal and opposite reaction and all that.) Air is pushed down, some of the reactive force is exerted upward, and lift is produced. It's the "barn door," and it could lift the airplane, but you'd need a ten-mile runway to get the thing off the ground. The high speed required to maintain flight would also make for spectacular landings.

There's some reaction lift at work all the time on every airplane—even the belly of the bird contributes something—but Bernoulli's discovery has more significance for the student of light training-type aircraft. Bernoulli noted that whenever air moved over a curved surface, the pressure on that surface was reduced somewhat, and the faster the air moved, the more the pressure went down. With natural forces constantly moving toward equilibrium, the higher-pressure air below the airfoil tried to push the curved surface (airfoil) into the low-pressure area above, producing a lifting force. You can demonstrate the principle yourself. Hold a piece of paper so that it hangs down and away from you and looks like the top of an airplane wing. Now blow gently across the curved upper surface and watch the paper rise.

In theory, then, an airfoil shape moved through the air rapidly enough to produce more pounds of lifting than the airplane weighs should make it fly. In practice, you'll combine reaction lift with Bernoulli's lift by raising the nose of the airplane on the takeoff run and holding it there during the climb. In level flight, when airspeed has increased, there's more of Bernoulli's lift at work than of the reaction type—it's a matter of using what's available to get the job done.

Since lift is so important to the operation of the airplane, the methods at your disposal to alter that force are important. Simply put, whenever lift is greater than its opposing force,

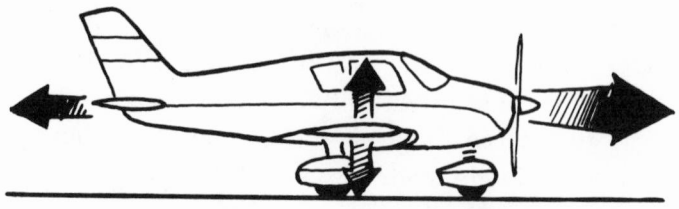

TAKE-OFF...ACCELERATING

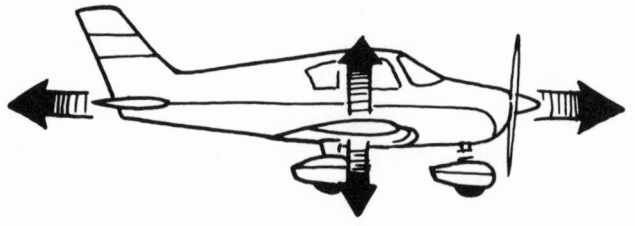

ALL FOUR IN BALANCE

Figure 1 The flight of an airplane depends on the

weight, the airplane will move upward; whenever lift is less than weight, the machine will descend. Lift can be increased by (1) increasing the angle at which the airfoil meets the air, (2) increasing the speed of its passage through the air, (3) increasing the amount of wing surface over which the air must pass (this can be done with certain types of wing flaps), or (4) increasing the curvature of the airfoil (again, with sophisticated lift-augmentation devices not generally found on training aircraft). Lift can be decreased by applying the opposite of

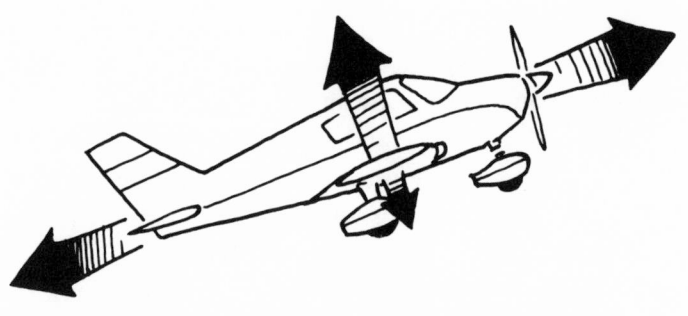

CLIMB... CONSTANT SPEED

POWER-OFF... DESCENDING

pilot's management of the four aerodynamic forces

any one or all of these arrangements. Any time the effective weight is changed, the relationship of lift and weight changes too, and the stronger force will always prevail.

VARIATIONS ON A THEME

Each flight is an original composition conducted by a pilot who is also the arranger for his own Four Forces Quartet. You'll be assigned a theme called straight and level, unaccelerated flight (which, by the way, is the most difficult of all in-flight maneuvers—you're trying to maintain lift, thrust, weight, and drag in perfect balance) and sooner or later be expected to come up with the variations: level turns, straight-ahead climbs and descents, turning climbs and descents, changes in airspeed. There's really nothing more you can do with an airplane; you can increase or decrease the size of the excursions from straight and level, but however you flop around the sky (even doing intentional aerobatics), you can't get away from those basic maneuvers. The relation of the four forces can be analyzed throughout a flight, starting right from scratch.

At rest on the ramp, there's no lift, no drag, and no thrust. The airplane just sits there, doing nothing, but weighing whatever it weighs. Start the engine and some thrust is developed, but not enough to overcome inertia. As soon as you get the prop going fast enough to throw enough air backward to produce a reaction large enough to overcome inertia, the airplane begins to move. (The propeller is nothing more than a rotating airfoil producing lift in a horizontal, forward direction.)

On the runway, power is increased until thrust again overcomes inertia, and as soon as air starts moving across the wings, some lift is developed. Drag begins to build up too, but it shouldn't equal thrust for a long time. At some point in the takeoff run, raise the nose a bit to let reaction lift offer Bernoulli's a hand, and when the combination is greater than weight, you're flying.

The designer has figured out the speed at which your airplane will climb best, considering the trade-offs of lift and drag. When that speed has been reached (note that thrust must be greater than drag for the plane to accelerate to that speed), most light airplanes continue the climb at full throttle, which means that the angle at which the wings cleave the air can be increased, improving the lift situation and keeping the thrust-drag relationship equal to prevent further acceleration.

Figure 2

Coordination—the proper amount of bank for the rate of turn being accomplished—results in a comfortable and efficient change of direction

When you've reached the desired altitude, you can stop climbing by either reducing thrust or lowering the nose. In the first case, drag will effectively increase, slowing the airplane and thereby reducing lift. If your objective is to fly level at the same speed used during the climb, some of the thrust that was being used to produce lift will be redirected forward as you reduce power. At the power setting at which all four forces are in balance, you will have accomplished your task. Notice that three items remained constant throughout this exercise: the angle of the wings, the airspeed, and the altitude.

If you lower the nose to level off, with no change in power setting, the airplane will accelerate, all the while maintaining the desired altitude. Now the constants change; with no reduction in thrust, the only way to prevent further climb is to lower the nose, reducing the angle at which the wings are moving through the air to reduce lift. Lowered nose means less drag, which has the same effect as increasing thrust: Speed increases (lift will grow—don't forget to lower the nose some more) until you reach the airspeed you want. On target, reduce thrust until it balances drag, and once again the four forces are in equilibrium.

Now to this matter of turning an airplane, and doing it right. A smooth, comfortable turn is the result of managing the four forces properly. Once you understand what is supposed to happen, you're more likely to *make* it happen.

Since the airplane is free to move in accordance with the forces acting on it, merely pointing the nose in a new direction won't get the job done. Oh, the airplane will eventually turn the way you point it, but it will be a reluctant, sloppy, and uncomfortable maneuver. An airplane flying straight ahead will try to keep on flying straight ahead, even if you point its nose another way. In other words, it'll fly a bit sideways for a while, until the forces all balance again. You feel the reaction in a car whenever you go around a curve a bit fast—the car and everything that's in it try to skid toward the outside of the curve.

The culprit is centrifugal force, limited to acting horizontally when you turn a two-dimensional vehicle like a car—unless the roadway is banked, as on a high-speed racetrack. On a banked roadway, the force trying to slew the car off the track is expressed against the sloping surface, and the driver feels himself being pushed down in the seat (increased effective weight). Race drivers are unable to control the amount of bank on the curves, but we pilots can make our machines do anything we want. So set up an invisible "banked track" whenever you wish to change the direction of flight, and let the people and things in your airplane be pressed gently, uniformly, and comfortably against it.

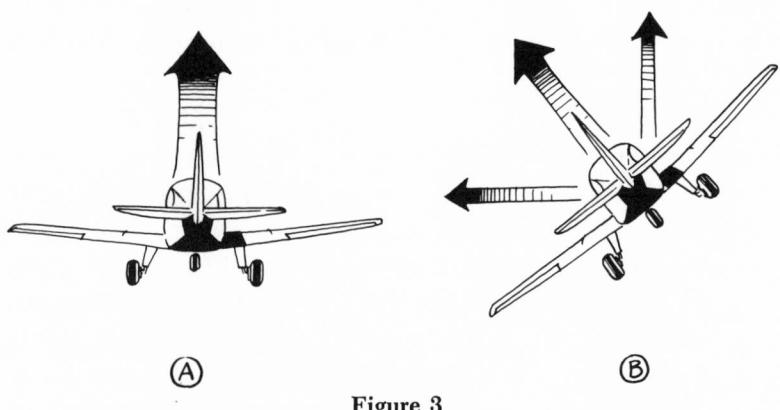

Figure 3

Vertical lift in straight flight (A), and its redistribution
during a level turn (B)

Come around behind the airplane now, and view the forces
of lift and weight from a new angle. In level, straight-ahead
flight (Figure 3A), all the lift force is acting in a line perpendicular to the wings, exactly equal and opposite to weight.
When the wings (the racetrack) are banked, some of that
vertical lifting force acts in the direction of the bank, literally
pulling the airplane around the turn (Figure 3B). It's the same
force, acting in the opposite direction, that forces the car to
turn as it follows the track. Instead of being pushed down by
centrifugal force, an airplane overcomes centrifugal force with
some of its lift. The steeper the bank, the more rapid the turn.

But you've disturbed the balance of the forces by redirecting
some of the lifting force, and the piper must be paid. With a
portion of the vertical lift now pulling you round the bend,
there's not as much force available to maintain altitude, so the
airplane will usually descend. On top of that, a banked turn
increases effective weight because of centrifugal force, so there
are *two* factors working to pull the plane downward. In order
to maintain altitude while turning, you've got to increase lift

by the exact amount lost, and there are two methods available: increase speed to get more air moving over the wings (that can be accomplished with more thrust) or increase the angle of wings to airstream to create more lift (or a combination of the two, depending on what you want to hold constant, airspeed or power). Of course, drag will increase if you choose to raise the nose for more lift, resulting in a lower airspeed. Can you get into a corner where there's not enough of *anything* available to keep the airplane on the desired altitude? A lot of pilots have found out the hard way that a steeply banked turn costs plenty in both lift and drag, especially close to the ground, when descending is definitely not the best thing to do in the interest of longevity.

You may have realized that one of the fastest ways to lose altitude in a hurry is to put the airplane into a steep bank without applying more thrust or increasing the wing angle. An inverted airplane, with the lifting force acting 100 percent downward, heads homeward in a hurry—but leave that to the fighter jockeys and air-show pilots. There's no need for more than 45 degrees of bank during your training days.

After all that, the management of the four forces to produce a descent should be clear. You have your choice of reducing lift by (1) reducing thrust, (2) reducing the angle of wing to airstream, (3) increasing weight, (4) increasing drag, thereby effectively reducing thrust, which reduces airspeed, which reduces lift . . . the combinations are endless.

There's one more aerodynamic property that needs to be explained. Not really a force, it's more a *lack* of force; in any event, it's very poorly named: the stall. To a society that's grown up with automobiles and motorbikes and the like, a stall means that the engine has quit; this is not at all what it was intended to describe to pilots. An *aerodynamic stall* is any condition of flight in which the wings are simply not producing enough lift to sustain the weight of the airplane. When this happens, weight takes over and the airplane loses some altitude until lift is once again produced in sufficient quantity to overcome weight, and all's well again.

The problem occurs if you're near the ground when the stall occurs. For the most beautiful landings of all, see to it that you run out of lift when the wheels are a hairsbreadth above the runway; for the most spectacular landings of all, run out of lift just far enough above the ground so that you won't have time to rebuild that precious lift before you run out of altitude. You probably won't want to practice *that* again.

There are as many ways to stall an airplane as there are pilots, but no matter how you approach it, the stall will occur when a particular aerodynamic situation is achieved. It's known as *exceeding the critical angle of attack*, and deserves some talking about.

Up to this point, angle of attack has been running around disguised as "the wing angle," or "the angle at which the wings cleave the air," or whatever; now you need something more definitive. An airfoil can be mounted in a wind tunnel, on a helicopter, bolted to an airplane, or swung around your head on a string—if it's truly an airfoil, it will produce some lift whenever air moves over its surfaces, *so long as that air moves smoothly*.

The wind-tunnel illustration is best, because you can easily imagine smoke streams flowing through the tunnel to provide visual clues to the behavior of air flowing past the wing. Stream lines represent the "relative wind"—the wing could be going straight up or straight down, but so far as it's concerned, the air it has to work with is coming from somewhere straight ahead.

An imaginary line from the leading edge to the trailing edge is called the wing's chord line, and furnishes a convenient place to measure the angle of the wing relative to the flow of air. This is the important angle, the *angle of attack*. In the wind

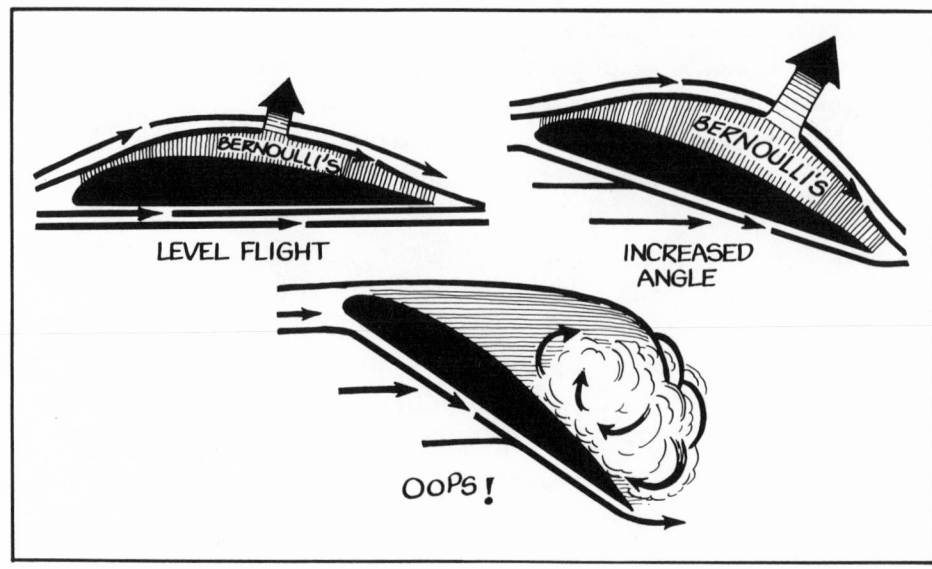

Figure 4

Regardless of the airplane's attitude, an aerodynamic stall
will occur whenever the critical angle of attack is exceeded

tunnel, the airflow remains fixed, and the airfoil is moved to
change the angle of attack.

Turn on the smoke, and notice that all the air passing *over*
the wing somehow arrives at the trailing edge at the same time
as the air that traveled *under* the wing. Since the top is curved,
there's no way that can happen unless the over-the-top air
moves faster, which it does—there's Bernoulli's discovery in
smoke. So long as the air moves smoothly, the top air continues
to meet the bottom air at the trailing edge; lift is produced in
direct proportion to the speed of the air.

When the angle of attack is increased, the air must travel
even faster to make the rendezvous. At some angle the distance
becomes too great—"burbling," or separation of the stream
lines, takes place, air must rush in from elsewhere to fill the

void, and lift is lost. When lift deteriorates to a value of less than the weight of the airplane, the stall happens. For most light aircraft, this critical angle of attack is in the neighborhood of 18 to 20 degrees.

That number is useless to you as a pilot, since there's nothing on the instrument panel that tells you when you're almost there, but boy oh boy, does the airplane itself let you know that it's about to quit flying! Horns sound, lights flash, the whole plane shudders and buffets—there's no mistaking the onset of a stall, and your instructor will have you practice it until he's certain you can't mistake it. Today's wings are designed to stall progressively, from the wing root to the tip, and with plenty of advance notice; the plane won't suddenly drop like a grand piano.

Recognition of and recovery from the stall is something every pilot must learn, because there's only one way out (reduce the angle of attack below the critical angle), and the price for recovery is almost always altitude. If you're high enough when the stall takes place, no problem—but if the ground interferes with your recovery procedure, big problem.

"STABLE" MEANS MORE THAN A HOUSE FOR YOUR HORSE

Ever watch a barn swallow going about his business on a summer evening? His scimitar wings and split tail let him change course and speed faster than you can watch. When your business is chasing bugs, you'd better be able to turn on a dime if you don't want to go hungry.

Then there's the pelican, the bird with the unfavorable beak-to-belly ratio, who tucks up his feet and loafs along in sedate flight punctuated by an occasional controlled crash into the sea when he spots a fish.

Each of these creatures possesses an aerodynamic quality

known as *stability*: the tendency of a flying machine to maintain a given condition of flight, and to return to that condition after its attitude has been disturbed. A swallow, on the low end of the stability scale, finds it very easy to make a sudden change in direction or speed of flight, because there are few aerodynamic forces at work to impede that change. A pelican is configured for low-and-slow cruising flight; when he needs to execute a sudden and rapid descent for his dinner, his broad wings and tail leave him no choice but to fold everything up and assume the flight characteristics of a brick.

General-purpose aircraft designers try to put their products somewhere in between, with enough stability to return to a safe condition of flight with no input from the pilot (a safety concern), yet not so much that they're difficult to maneuver (a creature-comfort consideration). In aviation's infancy, when stability was equated with quality of design, one fellow put together an airplane that was so stable it couldn't be turned. Unfortunately, it was aimed at a barn on its first (and last) test flight.

The criteria for stability in aircraft design fill volumes. It's enough to know that general-purpose aircraft must demonstrate the ability to maintain straight and level flight, and to return to straight and level flight within certain limits of time and altitude. Within its design limits, then, your airplane should fly quite nicely all by itself when it's properly trimmed, and should find its way back to straight and level flight sooner or later whenever you point it in some other direction and turn it loose. By understanding the design features that promote stability, you can aid and abet the forces at work, and become a smoother pilot. Stability is a three-dimensional, interdependent thing. When you change the airplane's attitude in one dimension there will be an effect on the other two—but it's better to learn about them one at a time.

Directional stability (Figure 5A) is first because it's the

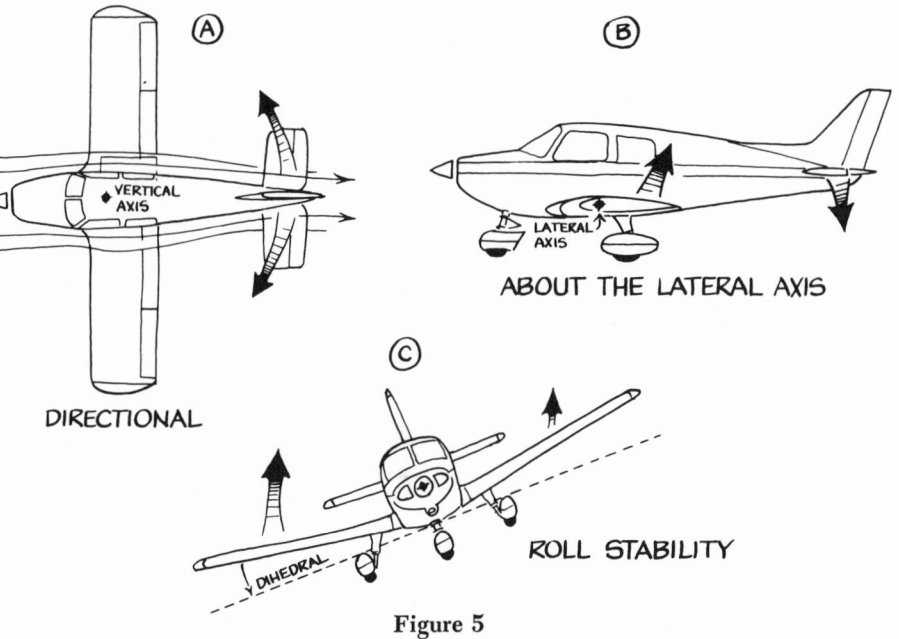

Figure 5

Aerodynamic forces incorporated in the aircraft's design exert pressures to return the airplane to straight and level flight

simplest in operation, therefore the easiest to understand. An airplane is nothing more than a big weathervane—it aligns itself with the relative wind because of the vertical stabilizer.

When the airplane is flying straight ahead, the relative wind moves equally over both sides of this airfoil-shaped fin, and no lift is created. But should the nose move to left or right around its vertical axis (that movement is called *yaw*), the vertical fin will move in relation to the air blowing past it, and lift will be created. The net result of this combined reaction lift and Bernoulli lift is increased aerodynamic pressure, which moves the aircraft's nose back toward alignment with the relative wind. The pressure acts smoothly—it doesn't snap the nose back— and constantly, always trying to correct any excursions from straight-ahead. When large displacements are made in the aircraft's heading, the reaction from the vertical stabilizer will probably be too strong and the nose will go through straight-ahead and a little beyond in the other direction—but not as far as in the first displacement. Now there will be a push the other way, but not as much, and the nose may swing back and forth a couple of times before the yaw is damped, or stabilized.

The behavior of an airplane when it is rotated about its *lateral* axis (an imaginary pivot running from one wingtip to the other) involves longitudinal stability, and the horizontal stabilizer (Figure 5B). Unlike its vertical companion, which seeks equal aerodynamic pressures, the horizontal stabilizer exerts a constant downward force to keep the airplane flying level. It's part of the price of lift—the center of pressure (that point where all the lifting force apparently acts) under the main wings is behind the wingtip-to-wingtip pivot, which means that the nose will constantly try to go down. Enter the horizontal stabilizer, an upside-down wing designed to provide a downward force just equal to the nose-down force in normal flight. It's particularly sensitive to changes in speed (faster air movement means more lift); when the entire airplane goes faster, the horizontal stabilizer pushes down harder to counter-act the increased nose-down force from the wings.

Put simply, every airplane has one speed at which these forces are exactly balanced; any excursions from that speed cause the horizontal stabilizer to attempt an equalization. For example, raising the nose causes a decrease in airspeed, which lessens the downward force of the stabilizer and allows the nose-down tendency to take over. Going downhill now, the airplane picks up a bit of speed, and the stabilizer becomes more effective and returns the airplane to level flight. The op-posite sequence takes place if the nose is depressed at the start. Just as in yaw damping, the airplane may go through several oscillations up and down before it settles back into a state of equilibrium, but each is smaller than the one before. Most light aircraft stabilize after three or four pitch oscillations.

Roll stability, or banking stability, is the product of two design features in most light airplanes, dihedral and keel effect, neither of which operates very effectively in a perfectly co-ordinated turn. But since you're concerned about the airplane's tendency to return to straight and level flight by itself, lower a wing and let go (Figure 5C). There will be some slipping, or movement toward the inside of the turn—gravity, you know— and keel effect will come into play. It's simple in concept: Most of the weight of the airplane is concentrated low in the fuselage, and any side force produced as a result of the slip tends to right the machine.

Dihedral is a bit more involved. It relates to the angle, viewed from ahead of or behind the airplane, at which the wings are bolted on at the factory, and varies among manu-facturers from barely noticeable to apparently ridiculous—but each configuration was designed for a specific use, or repre-sents a compromise. Whenever a dihedral angle is built in, the "down" wing presents a greater angle to whatever relative wind might be produced by the sideslip. In this case, relative wind from the side (a very small component of the total) cre-ates additional lift on the down wing and tends to push it upward, trying to return the airplane to a wings-level attitude. In most light aircraft, keel effect and dihedral are relatively small forces, meaning that those airplanes are relatively easy to bank, but relatively slow to recover from rolling displacements.

Putting it all together, stability is the *feel* of an airplane. Transport airplanes usually feel as though they're on railroad tracks—very stable, positive, ponderous. Aerobatic airplanes and fighter airplanes often feel as though they'll go charging off in some other direction if you blink your eye—and they often do! Stability can be whatever the designer wants it to be, but for everyday flying, there's a comfortable, safe compromise that will keep the airplane on an even keel when you divert your attention to other things and will help you get your aeronautical ducks back in a row when things go sour.

The means by which a pilot exerts his will over the flight path of his airplane has changed precious little over the years. Systems are more refined, more sophisticated, and easier to operate, but the results are the same as when the brothers Wright lay on the bottom wing and used a hip cradle for roll control. When the prone position fell from fashion, aircraft designers supplied yokes for sitting-up pilots; no doubt under the strong influence of a bicycle-oriented society, they thought the natural leaning motion that carried a two-wheeler around a curve would apply equally to aircraft. Leaning into a turn (or away from it) is a no-no these days. Control of all three dimensions of flight is vested in a control stick or wheel and two rudder pedals, which require only pressure, not movement, to make the airplane do your bidding.

Aircraft control deals with two principles of aerodynamics: movement of the airplane around its three axes and the production or destruction of lift. Control surfaces are devices that augment the natural stability of the airplane. If you were satisfied to always move through the air in straight and level unaccelerated flight, there'd be no need for controls, but it would be difficult to take off and land. By altering the pressures exerted by fixed stabilizers (for now, consider the wing the biggest fixed stabilizer of them all), and control surfaces—devices that augment designed-in stability—the pilot can move the airplane in the three dimensions of flight, singly or in combination. As before, one dimension at a time:

Yaw is controlled by the *rudder*, a hinged surface on the trailing edge of the fixed vertical stabilizer (Figure 6A). Strong cables or push rods connect the rudder to the pedals on the flight deck so that when you press on the right pedal, the rudder is deflected to the pilot's right. In straight flight, air pressure is equal on both sides of the airfoil-shaped vertical

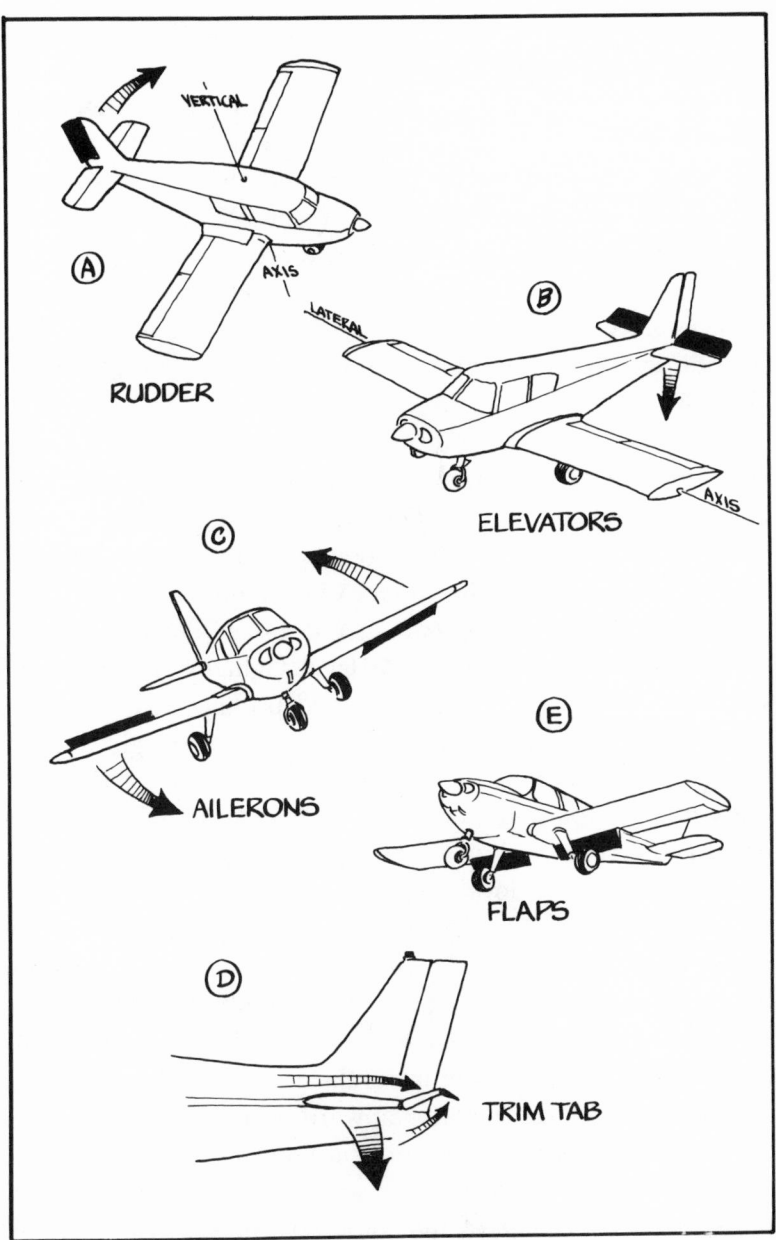

Figure 6   Flight controls

stabilizer, but when the rudder is moved, the air moving around the left side must speed up if it's to meet the right-side air at the trailing edge. Increased airflow means decreased pressure on that side (Bernoulli) plus some reaction from the air being displaced by the rudder, and the result is a push against the right side of the weathervane. When the tail moves to the left, the nose must move to the right, creating motion about the vertical axis—yaw, or heading change. Most rudder controls are very sensitive; you'll seldom if ever need more than smooth, even pressure on the rudder pedals of a light-engine airplane.

In general, yaw is a condition to be avoided except in certain very specialized situations. Why? Mostly because it's an uncomfortable condition of flight, with the airplane and everything inside it trying to move sideways. Your efforts should be concentrated on using the rudder to overcome unwanted yaw.

The next dimension—*pitch*—is the up-and-down movement of the nose, made to behave in level flight by the horizontal stabilizer, and controlled by a hinged trailing edge known as the *elevator* (Figure 6B). (Many of today's light aircraft incorporate a *stabilator* or *flying tail*, a blending of the functions of elevator and horizontal stabilizer in one unit. The effect is the same, and you can't tell the difference when you're at the wheel.) In level flight, you'll recall, the horizontal stabilizer produces enough downward force to hold up the nose—which is fine so long as there's no need to do anything but fly level. But you can change the magnitude of that down-lift by operating the control column or stick fore and aft, pulling on the cables or rods to move the elevators. Aft pressure on the wheel moves the elevator upward, forcing the underneath air to move faster, and you can figure out the rest of the story: more downlift, tail moves down and nose moves up. Forward pressure reduces some of the stabilizer's downlift, the aircraft accedes to the greater force, and the nose moves lower.

There's a lot of built-in mechanical advantage and a lot of aerodynamic leverage in those barn-door elevators way back there at the tailmost extremity of the airplane; if you really put your shoulder to the control wheel, you can turn your flying machine inside out (*not* a recommended procedure). Like rudders, the elevators thrive on even, smooth pressure, and passengers will love you for it.

Most complicated of the three, perhaps, but only because there's asymmetry involved, are the *ailerons*, aerodynamic surfaces that move the airplane into or out of a bank—the dimension of *roll* (Figure 6C). Ailerons are hinged to the trailing edges of the wings and are operated in one of two ways—by turning the *control wheel*, or by right-and-left movement of the *control stick*. The results are identical: Control pressure moves the ailerons in opposite directions—one up, the other down—to achieve the desired effect.

"Aileron," one of many words in the aviation vocabulary bequeathed to us by the French (fuselage, empennage, longeron, touche-et-go), means "little wing." Lift augmenters in the truest sense, ailerons generate an imbalance in the lifting force exerted by the wings—since the airplane is free to roll about the axis running from nose to tail (the longitudinal axis) whenever there is more lift on one side than the other, it will. So, to begin a bank to the right (looking from behind the airplane), control pressure applied to the right moves the right aileron up, the left one down. Over-the-top-of-the-wing air on the left side must move farther, therefore faster, and more lift is produced; at the same time, the right aileron has moved upward, encouraging the air flowing around that portion of the wing to move faster underneath the wing, creating some downlift. With increased upward force on the left wing and increased downward force on the right wing, the airplane banks to the right. In the absence of any other factors, a bank to the right will result in a turn to the right; some of the lift that was acting straight up is now acting horizontally, pulling the aircraft around in a curved path.

Unfortunately but unavoidably, the four forces get into the act whenever the ailerons are operated, and particularly drag, which is always greater on the high-pressure side of an airfoil (the lowered pressure on top of the wing, necessary to create Bernoulli's lift, also reduces the density of that air, which in turn generates less drag). Apply aileron-control pressure to the right again, and see how the control surface that is moved into the higher-pressure air (under the wing) of necessity creates more drag than the aileron that is moved upward into low-pressure air; the left aileron is "held back" more than its opposite number. The airplane, free to move about all three axes, makes up for this asymmetrical drag by yawing to the left— but you wanted to turn to the right! That's *adverse yaw*, as undesirable as just plain yaw, and equally uncomfortable.

You've two ways to eliminate or prevent yaw—use the rudder (so *that's* why it's there!) or buy an airplane with Frisé ailerons. To exercise the first option, know that the nose will try to turn left when you bank to the right, anticipate what's going to happen, and as you roll into the bank apply just enough right-rudder pressure to counteract the yawing. A Frisé aileron is a clever arrangement that deflects the down aileron a little less than it moves up the aileron. Less surface moving into the high-pressure air means less drag, so when you roll a Frisé-rigged airplane, it rolls—period. You'll be pleased to discover that most contemporary airplanes are so equipped, and you can make smooth, coordinated turns all day long with your feet flat on the floor. *Vive la différence!*

It should be clear by now that operation of any or all of the primary controls (rudder, elevators, and ailerons) has some effect on the attitude of the aircraft, and when the attitude is disturbed about any of the axes there are forces at work constantly trying to return the plane to its former position. A strong tendency toward stability is built into trainers and nearly all general-purpose airplanes, and that's good for safety's sake—if all else fails, turn loose, and the airplane will return to straight and level flight sooner or later. But on many occasions you'll want to maintain some attitude other than straight and level, which means you must hold the controls against those constant stabilizing forces. Enter the secondary control surface, known as a *trim tab,* and exit the requirement for the pilot to exert continuous pressure to keep the airplane pointed in the right direction.

In a normal climb, starting from level flight, the nose must be raised and thrust increased, and once everything settles down in the new attitude, the climb will continue—so long as you hold the nose up with back pressure on the control column. Let go and the stabilizers will take over, lowering the nose until the stability airspeed is reached. Holding that elevator pressure becomes very uncomfortable, and difficult, after a while. A trim tab, hinged to the elevator's trailing edge, can be moved in the opposite direction until *its* aerodynamic force is sufficient to hold the elevator where you want it (Figure 6D). In effect, you have established a new stability airspeed so far as the elevators are concerned.

Some trim tabs are controlled with hand cranks, some with levers, and some with electrical switches, but it matters not how they're operated—they all do the same job. And, like the other controls, they are to be operated with smooth, even pressure. Whenever you need to maintain pressure on a trim-tabbed control, ease the trim wheel or lever or switch in the appropriate direction until the pressure on the primary control lightens and disappears. Whenever you make a change in attitude or thrust or drag, you'll have to retrim or sit there holding the controls against aerodynamic pressure. A fully equipped airplane will have tabs on all three primary controls, but don't expect anything other than elevator trim on a lightweight trainer.

The fundamental laws of aerodynamics will not be denied;

somehow, any airplane that is banked gets that way because of redistribution of the lifting force. The Wright brothers accomplished the rolling motion by actually warping or bending the wings of their *Flyer;* greater angle of attack on one side resulted in more lift on that side, and the whole machine banked. It's not unlikely that more and more airplanes will adopt the "spoiler" technique that's been in use on gliders and sailplanes for years; such a plane, absolutely devoid of ailerons, has smooth wing surfaces until the pilot applies pressure to roll into a turn, at which time flush-mounted units rise up to partially disturb or "spoil" the smooth airflow over the top of the wing. The airplane doesn't care how it got that way, but it knows that there's more lift on one side than on the other, and it can do nothing but roll toward the low-lift side.

And of course you've seen the famous Beechcraft Bonanza with its "V" tail—where's the rudder? Walter Beech concluded that if he could build an airplane that didn't have to drag three stabilizers through the air, it would go faster or use less power at a given speed—hence the *elevon* or *ruddevator* (your choice —either verbal corruption tells the story). When a V-tail pilot operates the wheel back and forth, the elevons move up and down together; when he presses on the pedals, the ruddevator panels move more on one side than on the other. The rigging must be straight out of Rube Goldberg, but it works.

You might question the whys and wherefores of a delta-winged airplane—the vertical stabilizer and rudder are apparent, but are those control surfaces along the trailing edge of the wings ailerons or elevators? The answer is they're both, acting in concert to produce pitch changes, in opposition to make the airplane roll, or in any combination of the two.

There are as many other ways to augment the four forces and change aerodynamic characteristics as there are aeronautical engineers and needs for specialized airplanes. One lift augmenter with which you'll probably become very familiar is the wing flap, a fancy board that can be extended from the trailing edge of the wing (Figure 6E). Extended partway, it increases the distance that the upper-wing air must travel, hence more lift—a bit more lift than the omnipresent drag it also creates. As the flap is moved farther into the airstream, drag builds up until the flap becomes an air brake. In the light-aircraft world, wing flaps earn their keep mostly by helping to slow the airplane for landing.

Those are the very basic and fundamental concepts upon which flying machines operate: stability, control, and the four forces, all at your disposal as soon as you start a flight. How well you handle the airplane and how safely—above *all*, how safely—you fly depend largely on your knowledge of what's happening and your management of the immutable laws of aerodynamics.

Aircraft Performance

2

There's a simple fact of aviation life to be faced: Even though every airplane in the fleet comes equipped with charts and diagrams and graphs to cover every conceivable situation, most pilots play aircraft performance by ear. That's not all bad, because a pilot who flies the same airplane time after time will begin to recognize its limitations—he'll know when a runway is too short, a mountain too high, or when a proposed trip will require a fuel stop somewhere along the way. But the time-after-time process that leads to that knowledge can be peppered with some pretty harsh lessons. Learning how much runway an airplane requires by plunging off the end of one that's too short is not a good deal in anyone's book, nor is the uncomfortable realization that the hill ahead can't be flown over. Experience is without a doubt the best teacher, but it's also the most expensive.

For two reasons, the true professionals of aviation—airline captains, military aviators, commercial pilots—are great respecters of the numbers that apply to their airplanes. The laws of flight leave no room for guesswork about performance, and the types of airplanes these pilots fly show considerable variations in all parameters of flight when performance discipline is not maintained. The accident files are stuffed with evidence of the unhappy results when heavy-iron drivers attempt to make their machines fly too soon, not soon enough, or too fast or too slow. In general, airplanes with higher performance capabilities are less tolerant of pilot ham-handedness—a very fortunate state of affairs, given the nonprofessional abilities of the average general aviation pilot, and the fact that most noncommercial lightplane operations are conducted from airports with more than enough runway, over terrain and through weather that don't really test the pilot's skill. And that's the way it should be.

All of which is not intended to imply that it doesn't make any difference. Even the worst pilot can do better if he flies his airplane by the numbers; but detailed, precise performance planning takes time—and, quite frankly, unless the field is short, the terrain high, or the trip long (perhaps all three), that time cannot be justified for the average light-plane operator. *However,* learning by actual experience just what your airplane will or won't do is often learning the hard way, and your graduation from aviation's College of Hard Knocks can be hastened considerably by recognizing the situations that are going to cause problems. How much runway does your airplane need to get off the ground? How much more when it's loaded to the limit? When you have at least ballpark answers to these questions and others that bear on total performance, you can make sensible and safe decisions; what may be even more important, you'll know when to dig into the performance charts to get specific answers when it looks as though an upcoming flight operation may be a tad on the marginal side. When that situation comes along, you can bet your life (literally) that the time spent with the books will be justified. How sweet it is to bask in the knowledge that unless the airplane comes completely unglued, it will do what needs to be done; how terribly uncomfortable to realize suddenly that you're not going to clear the trees off the end of the runway, and that you're going too fast to stop before you get there.

ROCKET PILOTS MAY NOW BE EXCUSED

All flying machines that depend on air to provide thrust and lift must hang their performance hats on the density of that air. No matter how good its pilot or how impressive the sales-brochure statistics, an airplane's performance improves in thick air and gets worse as the air gets thinner. The range at any specific location (such as the airport from which you operate most of the time) is not usually a very wide one, and since one extreme —a high-density situation—can only improve performance and sweeten your flying experience, your concern should be to recognize and take into account the effects of low-density air.

Right off the reel, certain immutable laws of physics dictate the responses of a gas (air) to changes in temperature and pressure. If the gas is uncontained—that is, free to expand and contract—it will increase its volume when the temperature is increased or the pressure is decreased . . . and our atmosphere follows this law to the letter. In very general terms, you should expect lousy airplane performance on a day when the temperature is high and the barometer is low.

Temperature variations are easy to understand and recognize, but changes in pressure are not so easy to detect. Very seldom does the barometer venture more than a half inch above or below the normal 29.92 inches of mercury, and we have gotten so accustomed to these slight, slow changes in pressure that they aren't usually noticed. On top of that, the air-pressure changes that affect aircraft performance come from two sources: the weight of the atmosphere pressing against the surface of the earth and changing constantly as weather systems move along, and the altitude of the airplane. After all, we live at the bottom of an ocean of air, which means that each foot we swim upward decreases the pressure a little bit. Your search for the very worst performance situation ends at a high-altitude airport in the eye of a hurricane on a hot day.

Rather than calculate the effects of pressure and temperature on each contributor—engine power, thrust, and lift—the aviation community puts them all together and furnishes performance information for a specific situation, with corrections for abnormal atmospheric conditions. That specific situation is the pressure-temperature combination at sea level on a *standard day*, when a barometer reads 29.92 inches of mercury and thermometers stand at 15 degrees Celsius. ("Celsius" is the official name for the metric temperature scale, which is being adopted in nearly all aviation weather reporting. Only surface temperatures are given in degrees Fahrenheit.) Under those conditions, air density is constant and predictable; if standard-day conditions prevail throughout the atmosphere, each 1000 feet of altitude increase will result in a decrease of 1 inch of pressure and a drop of 2 degrees in temperature—not only does the air expand when the pressure is reduced, it gets cooler. These standard-day decreases are known as "lapse rates."

With the knowledge that pressure and temperature change constantly and predictably, air density can be calculated for any given height in a standard atmosphere—and, conversely, if you know those values (pressure and temperature) for the level at which you intend to operate your airplane, there's a chart that will provide the standard altitude represented by those conditions. For example: At a sea-level airport on a standard day, the pressure is 29.92 inches, the temperature is 15 degrees C., and the standard altitude is 0 feet; at 5000 feet above that airport, the pressure should be 24.92 inches (five times the lapse rate of 1 inch per 1000 feet) and the temperature should be 5 degrees C. (five times the lapse rate of 2 degrees per 1000 feet). The air would be as dense as 5000-foot air should be on a standard day, and your airplane will perform accordingly.

But suppose the pressure has dropped more than 1 inch per

1000 feet because of a weather system, and at 5000 feet the barometer reads 23.92 inches while the air has cooled to only 10 degrees C. Under these conditions the air at 5000 feet is less dense than it should be—its density is the same as the density at 6500 feet on a standard day, and the airplane will perform as if it were flying at 6500 feet. Aircraft performance is inescapably tied to the *standard-day altitude*, which means that whenever you know the pressure-temperature numbers for any level in the atmosphere and can get your hands on a chart or computer to determine what standard-day altitude that combination represents, you've computed a *performance altitude.** If it's higher than the actual altitude at which you're flying ("actual" in terms of feet above sea level), the performance will be somewhat less than you'd expect at the actual altitude. When the performance altitude is lower than the actual altitude, that's the kind of day you should choose to demonstrate an airplane to a prospective buyer—the machine will fairly leap off the ground and will climb like a rocket.

There are at least two ways to calculate performance altitude: with a navigational computer (set the *pressure altitude* opposite the outside air temperature on the appropriate scale and read the performance altitude where the computer directions tell you to look) or with a chart specifically designed for that purpose. Either way, you'll need a value for pressure altitude, which can be determined in the airplane by referencing the altimeter to 29.92 inches (that tells you how many feet you are above the standard pressure level) or by making the appropriate adjustment to field elevation based on altimeter setting. Keep in mind that atmospheric pressure must drop 1 full inch to raise the pressure altitude 1000 feet, a rare situation, and one usually accompanied by weather so foul you won't

---

**Performance altitude* is a Taylor-made term, by the way. You'll find *density altitude* and *standard altitude* elsewhere; although they are interchangeable, I think that performance altitude is more relevant.

want to go flying anyway, so unless there has been a significant drop in pressure, your performance-altitude ballpark number won't be very far off if you use field elevation or cruising altitude.

Temperature has a much greater effect on performance altitude, and fortunately there's something you can do about it— you can plan to fly in the early morning or late evening, when a lower temperature gives you some air to work with. Particularly in the high country, siestas also apply to airplanes.

There is another atmospheric problem that is ignored by the architects of light-aircraft performance charts. Relatively insignificant, but worth knowing about, the moisture content of the air has an adverse effect on aircraft performance. Air is a mixture of gases (nitrogen is by far the most prevalent), and when one part of the mixture increases its presence the others must take proportional back seats—as on a moist, muggy day, when there's a great deal of water vapor in the air, replacing the other gases with its super-light combination of two parts hydrogen (the lightest element of all) to one part oxygen. Every cubic foot of such moisture-lightened air is less dense, resulting in lowered overall aircraft performance. Since you don't have charts to tell you how much it's lowered, make it your practice to be a shade on the conservative side when the air is heavy with moisture—particularly on those lazy, hazy days of summer, when high temperature combines with high humidity to put you in a performance corner you might not be able to fly out of.

Aircraft builders have long been required to supply a complete set of performance charts with each airplane, but the manufacturers were apparently behind the door when the standardization rules were passed around, for each has his own way of presenting the information. One of these days all charts will look alike, thanks to a manufacturers' organization that realized the disservice created by a hodgepodge of formats; but for a long time to come the operators of America's flying machines are likely to encounter several considerably different methods of arriving at the performance numbers. One manufacturer asks the pilot to compute performance altitude separately, then apply other factors; another incorporates a chart which doesn't provide a direct readout of performance altitude, but is used to show the effect of temperature and pressure variations on performance; a third method uses a tabular format—standard-day numbers for frequently used altitudes, and adjustment factors to be applied for nonstandard temperatures; still another manufacturer's concept has a simple table with altitudes on the side, temperatures across the top—find the intersection of the lines that represent the conditions at hand and there's the performance number.

WEIGHT, WIND, SLOPE, AND WATER

No matter which type of chart or table you use, there are things other than performance altitude to be considered, such as aircraft weight, which will obviously cut down overall performance as the number of pounds increases. On takeoff, for example, the wings require a certain airspeed to generate enough lift, and since thrust will be the same regardless of weight, a heavier airplane will accelerate more slowly and use up more runway in the process of attaining takeoff speed. At cruise, weight becomes the arbiter of performance: At a given power setting a heavy airplane will fly slower than a light one, and if you decide to fly at the same airspeed come what may, you'll find that more power is required to make it happen with an airplane that weighs more. If you understand aerodynamics at all, you'll know why: At a given airspeed, the wings' angle of attack must be increased to produce more lift and thereby more drag, which requires more thrust if level flight is to be maintained.

You're on final approach to a short field with a heavy airplane now, and about to discover what kinetic energy is all about. Even at the slowest possible approach speed, every extra pound on board adds to the mass moving down the runway, a mass that comes to a stop only when all the energy of movement is somehow dissipated—through heat in the brakes, rolling friction, running the airplane through a hangar, or any one of several other effective but unpleasant alternatives.

On-the-runway performance—the distance to become airborne and the distance to touch down and stop—is very dependent on the wind. With all other factors constant, an airplane will get off the ground sooner and consume fewer feet of runway on landing when those two operations are conducted into the wind. On takeoff, it's a matter of the wings' sensing airflow even while they're standing still; airflow means lift, even if only a minute quantity, and whatever lift is generated by virtue of the wind is that much less to be developed from forward movement of the whole airplane. On the other end of a flight, an into-the-wind landing is a higher-performance landing, because the groundspeed and therefore kinetic energy are less. While a tailwind is pure joy on a cross-country trip, it is one of aviation's nasties for takeoff and landing; any air movement from behind means that your airplane must accelerate to normal takeoff speed plus the speed of the wind before the wings can provide sufficient lift to get everything off the ground—and more acceleration means more distance from brake release to liftoff. Kinetic energy on a downwind landing is increased considerably, because the wind from behind pushes the airplane across the ground, and the stopping distance grows accordingly.

Takeoff and landing performance is always improved by operating into the wind, hence the general rule that pilots leave the earth and arrive thereon using the runway most nearly aligned with the wind. There are occasions, however,

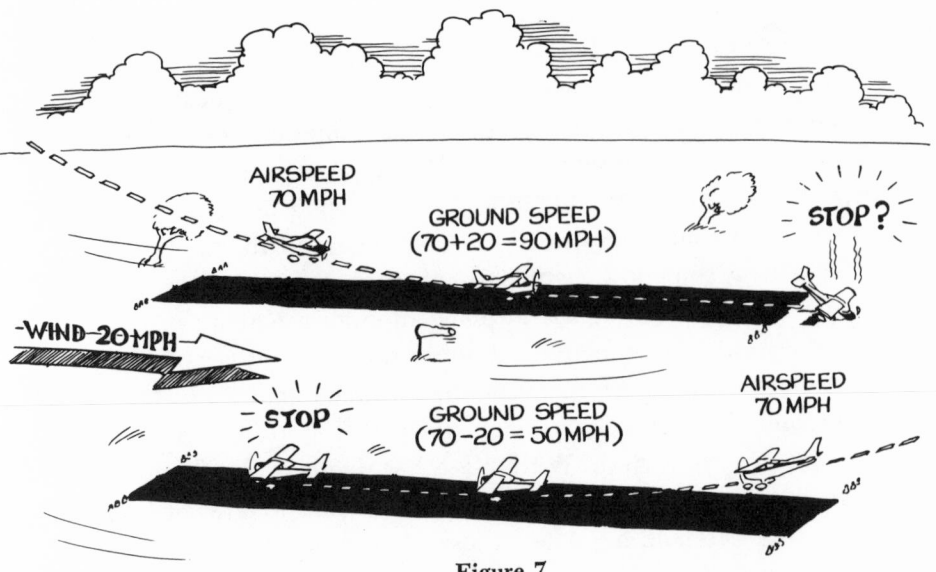

Figure 7

Aircraft performance (in terms of runway required to come to a stop after landing) is always derogated by a tailwind

when the zephyrs are moving about only lightly and perhaps coming from variable points of the compass; it seems a waste of time to taxi the length of the airport for the sole purpose of taking off into the wind when a downwind departure would present no performance problems (plenty of runway, light load, no obstacles to be cleared immediately after takeoff), and you would expect the chart builders to include a set of numbers showing the effect of going with the wind instead of against it. Curiously, you'll find such charts only in the books for the bigger airplanes. They're trying to tell you something: If you haven't enough experience to quantify the effect of a tailwind, consider all tailwinds bad news and make your take-offs and landings with the wind in your face. (Under certain conditions—heavy load, low-performance altitude, short runway, and so on—wind can be a significant factor in performance planning, and it gets more important as its velocity changes. However, a pilot proposing a flight whose success depends on a steady wind is a pilot inviting embarrassment at the least and courting disaster at the worst. Wind is perhaps the most fickle manifestation of Mother's capricious Nature; although the air may be moving in your favor at the start of the takeoff roll or as you turn onto final approach, it's as likely as not that the direction or velocity or both will change at that crucial moment when you need all the help you can get. Pilots shouldn't be superstitious, but there are little people called gremlins who sit in the sky watching for just such opportunities to turn the wind off or turn it around.)

The takeoff and landing performance charts might be considered a contract between you and the folks who built the airplane—you agree to supply normal pilot technique and the manufacturer supplies aircraft performance in consideration of all the factors of wind, weather, and weight. But the fine print in that contract specifies a level, hard-surface runway. What about a takeoff or landing on a sloping airport? Or one of those roller-coaster runways where the middle of the strip is considerably lower (or higher) than either end?

Faced with an even slope and no wind, you can put gravity to work for additional acceleration, so a downhill takeoff is definitely advised. When the wind is blowing down the slope, you'll have to make a judgment on the merits of downhill acceleration versus a tailwind takeoff. In the case of a sunken-center runway, you'll have help to the halfway point, but it's all uphill after that; a humpbacked airport will be just the opposite. You'll find no charts for runway slope in the light-aircraft performance books, so experience becomes the deciding factor. There are so many combinations of slope and wind (to say nothing of the trees, power lines, and rock walls that frequently jump up right off the end of these runways) that you'd need a library of charts to cover all the possibilities. And the non-level runways will most likely be located in the mountains, where the lay of the land leaves little choice, where altitude takes its toll of performance, and where the smartest thing to do is confer with the local pilots—you can profit from their experience. When in doubt, lighten your airplane—right down to bare bones—and give it a try. If all goes well with a light load, come back and try again with more pounds on board. You'll soon discover a weight you can live with.

Numbers from the performance charts are also invalid when the runway surface is anything except pavement. The wheel drag of even a dry, just-mowed grass runway will increase the takeoff distance somewhat, but let the grass grow a couple of inches, let the surface get soggy from heavy rains, and it's quite possible that the *available* power will fall short of the *required* power for accelerating to takeoff speed. As in the case of runway slope, a pilot judgment will have to be made, and once again it's experience that makes the difference. Long grass, snow, mud, and the like will probably increase the takeoff distance by 100 percent or more. When in doubt, try it with a light load, or mow the grass.

GOING UP . . . BUT HOW FAST?

Vertical speed—rate of climb—is one of those performance parameters touted by aircraft salesmen, but it's little more than nice-to-know for most pilots . . . until you find yourself in a situation where you've got to make the airplane climb just as rapidly as it's able to or wind up in the trees. For any given aircraft weight, there is an airspeed that will produce the maximum number of feet per minute, and a series of weight/speed combinations that represent normally anticipated loads is always included in the performance charts for light airplanes. These speeds will produce a rate of climb that is a compromise of vertical and horizontal movement; known as *best-rate-of-climb* speeds, they provide the most gain in altitude per unit of time.

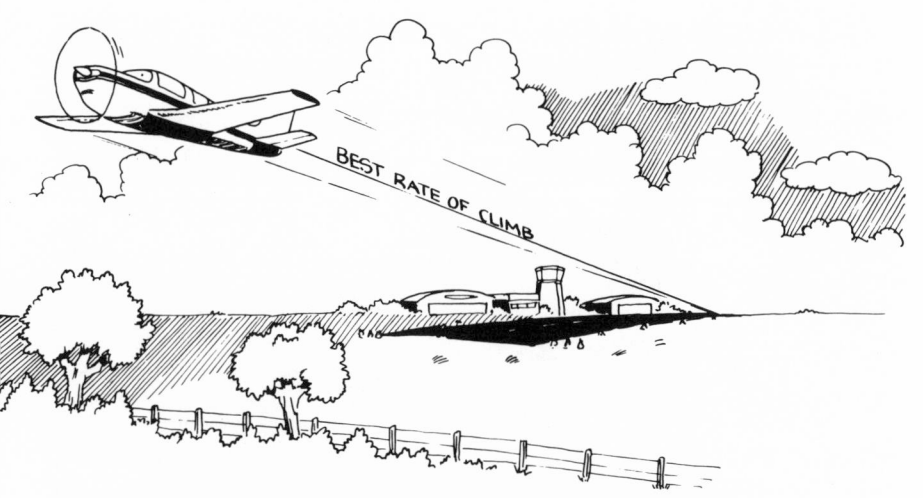

MOST GAIN IN ALTITUDE PER UNIT OF TIME
(FEET PER MINUTE)

Figure 8

Best rate of climb

On the other hand, absolute maximum climb performance can be achieved by increasing the angle of attack until the wing is just short of a stall, producing all the lift of which it's capable. The increased rate of climb that results is at the expense of airspeed, but when you need to climb over a tree *right now,* the angle of climb is a heck of a lot more important than your progress across the ground. The *best-angle-of-climb* speed (always a lower number than best-rate) will provide the most gain in altitude per unit of distance. Climb performance—best-angle and best-rate—presupposes a certain power setting, of course—usually wide open for fixed-pitch propellers and a specified climb power for airplanes equipped with constant-speed props.

In the charts for your airplane you'll likely find only one best-angle speed—it's intended to help you out of a tight spot close to the ground—but there will be a selection of best-rate speeds based on the two most effective overall factors: performance altitude and aircraft weight. Notice that the best-rate-of-climb

---

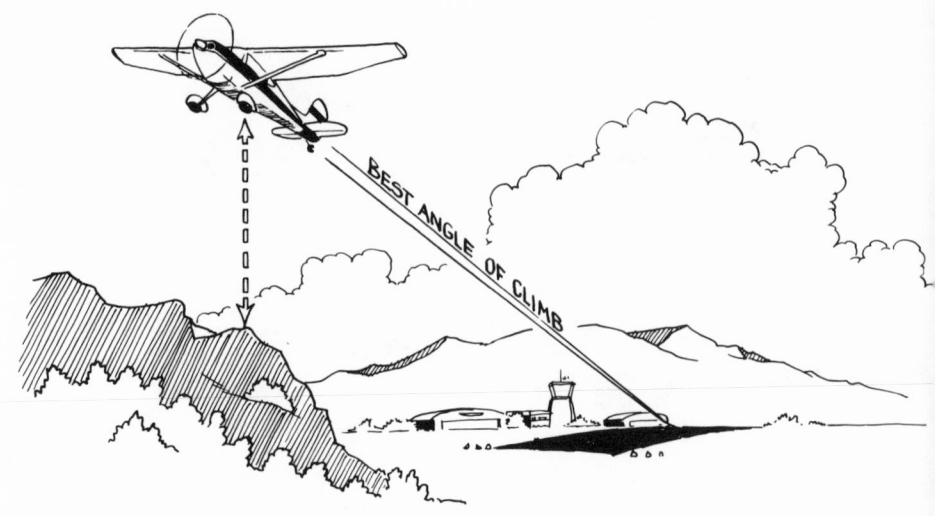

MOST GAIN IN ALTITUDE PER UNIT OF DISTANCE
(FEET PER MILE)

Figure 9

Best angle of climb

airspeed decreases with weight and altitude, and even though it's not necessarily indicated, chart builders are talking about performance altitudes. Don't expect to take off from Denver on a hot day and get the climb performance listed under the 5000-foot column in the charts—if the performance altitude at the Mile-High City is, say, 7500 feet, your airplane will climb at the rate specified for 7500 feet, no more.

As it turns out in the real world of flying, most pilots elect to climb at an airspeed somewhat higher than best-rate, a laudable practice because it bears on safety; the airspeed for best angle of climb requires a pitch attitude that will fill your forward vision with a great view of the aircraft engine—and even at best-rate airspeed you may feel uncomfortable about what you can't see out front. The problem will solve itself at high gross weights, when you'll have to hold the nose down significantly to obtain best-rate airspeed, and somewhere in between is a compromise attitude—and the speed to go with it—that

---

provides adequate forward vision and acceptable vertical progress.

The higher of the two, the best-rate-of-climb speed, finds application at the other end of the aircraft-performance spectrum. Because it's based on maximum efficiency of the wing, the best-rate-of-climb airspeed turns out to be very close to the *least-rate-of-descent* speed; there will be some slight changes that result from the lack of thrust, but when the engine quits you can glide farther at that airspeed. In other words, the wing is doing its very best aerodynamic work—and if your airplane has a lot of pounds between empty weight and maximum gross, look for a chart or table that specifies the best glide speed for different gross weights.

Depending on the type of flying you do, you may once in a while/frequently/never (cross out those that do not apply) come up against a situation where performance required and performance available are closer together than you'd like them to be. That's when the smart pilot gets out the charts to make sure—and if the gut feeling persists, a study of the numbers should bring him to the conclusion that weight is the most effective means of improving aircraft performance, all other factors remaining the same. In the final analysis, flying an airplane comes down to flying a wing. A wing can produce only a certain amount of lift; when there's less weight to pull off the ground or carry over a row of trees, the surplus lifting power can be used to improve performance—earlier liftoff, more rapid climb. On landing, less weight permits a lower stall speed and the slower touchdown that goes with it; it also means less kinetic energy, therefore less heat for the brakes to dissipate and fewer feet used up in the stopping process.

Fortunately, weight is also the factor over which you've the most control. Load planning (when an upcoming operation looks a bit shaky performance-wise) might take the form of partially filled fuel tanks instead of the usual "top 'em all," or informing your passengers (especially the fat ones) that you'll have to make one-at-a-time shuttle flights, or that unless everybody travels light you can't go as a group. Given the unforgiving nature of aviation, it's so much better to make an extra trip or stop along the way for fuel; once you're airborne and discover the airplane is too heavy to perform as it must, there's little that can be done.

THE FINAL FACTOR

Performance altitude is undeniable, weight is controllable, the runway slope is unchangeable, and wind effects can be determined right up to the moment the brakes are released, but the single element that probably has more to do with what actually happens in regard to aircraft performance is the one that is least predictable and varies more than all the others combined: pilot technique. Most performance-chart architects are realists, and leave a little room around the edges of the numbers to allow for flying skills that may be a bit rusty or were never really well developed; but remember that those numbers were obtained with expert pilots at the controls of new airplanes, with engines tuned like Maseratis at the Grand Prix. It's doubtful that the average pilot can achieve all the performance of which his airplane is capable, if for no other reason than that a reciprocating engine starts grinding on its insides and losing horsepower the very first time it's run, and continues downhill from there—a small loss, to be sure, but a performance theft nonetheless.

Given the overwhelming availability of comfortable-length paved runways, and the infrequency of occasions when the average pilot really needs to fly into or out of a short strip, there are two ways you can keep from painting yourself into a performance corner: If you must operate from a runway with little margin for error, get good at flying your airplane under those conditions (that requires training); when the amount of ground or sky that you'll need comes close to the amount available, remember that it's you versus the test pilot—add something for lack of experience and skill (that requires personal evaluation). And should the situation smack of impossibility, don't be there at all . . . *that* requires common sense.

So here's to performance:

May your runways always be long and slope away from you, your flying days cool and dry; and may the wind always be on somebody else's tail.

Aircraft Loading

3

On the door of the glove compartment in my car is a placard that states the weight limit in no uncertain terms—MAXIMUM LOAD 735 LBS.—followed by an equally clear admonition about the number of passengers: DESIGNATED SEATING CAPACITY—4 (2 in front, 2 in rear). Now, everyone knows a Volkswagen will hold more than that; the design limits presented a challenge for record chasers who insisted "there's always room for one more" until they had prodded, pushed, and packed 111 people into and onto a VW sedan. Even if they were all midgets, that's a lot of bodies in a Bug. Nobody expected the car to move with all that weight, mind you, but the record was set, and the Bug-stuffing champs went on to loftier pursuits (such as sardining people into telephone booths).

Airplanes have weight limits too, and although now and then you'll see an advertising photograph of an aviation company's newest product with the entire office force standing on the wings, there's a world of difference in the respect a pilot must show for the number of pounds and the number of people he puts in an airplane. Automobile manufacturers recommend weight and capacity limits, but airplanes have absolute maximums, backed up by the force of federal regulations.

Consider the obvious limitation first: Every occupant must be strapped to his seat, at least for takeoff and landing, which means that airplane's seating capacity is unequivocally limited to the number of seat belts installed; the only exception to the rule concerns youngsters less than two years old, who may be held by a properly belted adult. (Seat belts have been a way of life with aviators ever since someone noticed that pilots and passengers got hurt a lot less if they stayed with the airplane when a flight was terminated abruptly. The advent of shoulder harnesses in even the smallest of airplanes is long overdue and will no doubt save a lot of face—literally—in those unhappy situations when an occupant's countenance can be impressed on the instrument panel by the decelerative forces present in almost every airplane crash—but only if those shoulder harnesses are used for every takeoff and every landing. You just never know when an airplane is going to come to a suddener stop than the pilot had planned.)

While the seat belt–for–everybody rule is grounded in nothing more than concern for occupant survival, the legal limit on aircraft weight is directed more toward performance and maintenance of structural integrity—in other words, keeping all the pieces of the airplane together. An overloaded automobile may groan across the ground or collapse in a heap of broken springs, but an overloaded airplane can be put into an uncompromising situation when it's too heavy to fly and is moving too fast to be stopped on the pavement—hello, grass or trees or ditches or any number of other airplane-eaters off the end of the airfield. Occasionally there's enough lift to get the overloaded airplane off the ground but it refuses to climb; the sometimes-friend, sometimes-foe called *ground effect* provides a cushion of compressed air beneath the wings, which may be enough extra help to sustain flight—but it operates only within one wingspan of the ground, and who wants to fly an entire trip at an altitude of thirty feet? Suppose that somehow a pilot gets his too-heavy airplane off the ground, manages to climb

---

out of ground effect, and then must continue that climb or run into a mountain. If the engine is putting out 100 percent and the wings are generating all the lift they can and it's still not enough to clear whatever's in his way, the result is at best embarrassing, at worst tragic; and it can be traced right back to an easily committed sin—the pilot tried to fly an overloaded airplane.

Beyond the performance considerations, airplanes suffer from a limitation that car makers don't even know exists: *load factor*. The world-champion VW probably weighed 12,000 pounds with all those people in it, but that 6 tons just sat there; they didn't move forward or back, much less vertically. Even if the car could have been driven, the weight on its wheels would not have exceeded 12,000 pounds. But an airplane is frequently subjected to centrifugal force, which increases its weight. For example, when you roll your airplane into a 60-degree bank and apply enough back pressure on the controls to maintain altitude (in other words, turn in a level circle), the centrifugal force will be precisely the same as the actual weight of the airplane, and will add to the load the wings must carry. Your 2000-pound airplane suddenly weighs 4000 pounds so far as the wings are concerned. The *effective weight* in such a turn (or any other maneuver that imposes a similar load) is twice the actual weight, or a load factor of 2. You'll also see this force expressed in terms of units of gravity —G's—because any load at rest weighs whatever it weighs, with only the pull of gravity (1 G) acting on it. Whenever the load factor increases, the G-loading will increase with it, so a load factor of 2 is the same as 2 G's on the airplane and everything in it.

Except for those purposely designed for other than normal passenger-carrying chores, airplanes are built to withstand a load factor of 3.8, so you can wrassle that 2000-pounder around the sky, turn as sharply as you like, even play dive bomber—

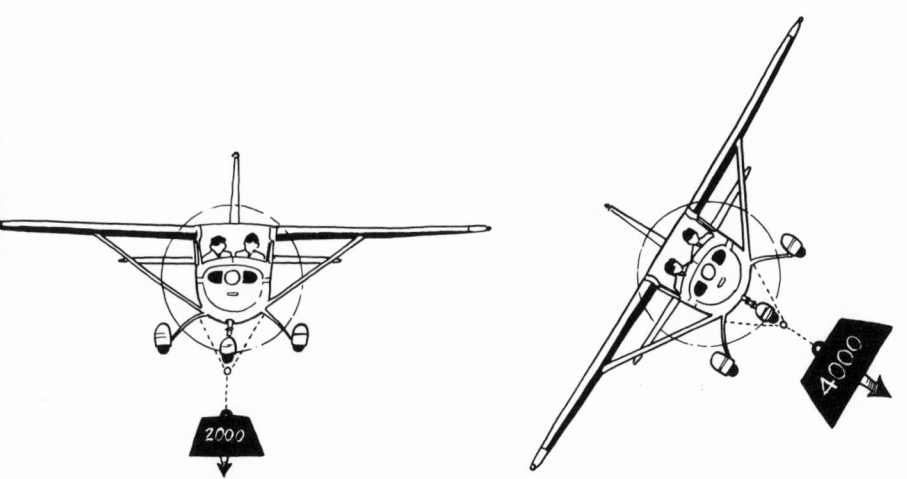

STRAIGHT AND LEVEL ... 1G   LEVEL TURN 60° BANK ... 2Gs

Figure 10

"G" force; the effect of centrifugal force in a turn, or pullout from a dive

the wings will stay with you up to a load of 7600 pounds; beyond that the manufacturer's guarantee evaporates. But consider the pilot who stuffs 2200 pounds into an airplane that has a 2000-pound weight limit. If 3.8 G's are imposed on this airplane, its wings are called upon to support, effectively, 8360 pounds. Two hundred pounds the pilot didn't think would make much difference become 760 pounds of additional load. The lesson is clear and simple: *Don't overload.*

NUMBERS YOU SHOULD KNOW

If an airplane were put on the scales just prior to takeoff, the dial would show the weight of the airplane itself plus the passengers, baggage, and fuel, and the pilot is responsible to see that this weight does not exceed the *maximum allowable gross weight,* or "max gross." Each airplane within a model designation (such as Cherokee 140, Cessna 182) has the same max gross limit, because the structural considerations are identical, but empty weight varies with the owner's preference for installed equipment. In general, airplanes come off the assembly line as barebones models at the same weight; then the customizing starts. Radios, extra fuel tanks, special seats, any number of optional extras are added that increase the price tag *and* the empty weight. That's the starting point for your weight computations, the number to which you add pounds of people and baggage and fuel. Empty weight is so important that it's specified in the airplane's papers and must be officially recomputed whenever it changes.

Between the empty weight and maximum gross weight is the number of pounds you have to work with: *useful load*. You can put together any combination of passengers, baggage, and fuel, but the useful load may never push the total weight over the top; you will always and ever be limited to the maximum allowable gross weight at takeoff. The combinations are endless and flexible, according to the purpose of the trip. In most light planes, when you need to carry a full load of passengers and baggage the fuel load will probably be limited; if range is important, somebody will have to leave a suitcase at home. Whatever the sacrifice, keep the gross weight under the limit. Especially in the case of lightweight airplanes that haven't stupendous useful-load figures to begin with, a lot of installed extra equipment (increased empty weight) gets pretty difficult to justify, unless you want to fly alone on short trips most of the time.

In practice, you need be concerned about only one of those three numbers. The empty weight is fixed, the maximum allowable gross weight can't be exceeded, so if you're careful never to put on board more than the useful load, you can't get in trouble weight-wise. It's probably only a matter of time until some clever fellow comes up with a low-cost electronic sensor to tell you exactly what your airplane weighs—on the ground, of course. While waiting for that little gem to filter down from the airlines and the military, where a similar system has been in use for some time, get acquainted with at least some ballpark weights for the load combinations you normally carry. When you have doubts—even little ones—about the legality of a particular load, total up the weights and make sure that what you propose putting into the airplane will fit between the empty weight and max gross.

FOREWARNED IS FOREARMED

There are only a few airplanes in the "light" class, up to and including the six-place twins, that are legal when the seats, the fuel tanks, and the baggage compartments are filled. In almost every case, such a load will exceed the allowable useful-load figure. When this happens, you—the pilot in command—will have to decide what's most important. If all the people need to go (and if you decide they don't, you're probably paying for more airplane than you need), explain the problem well ahead of time and ask them to travel light—no steamer trunks. If everybody is down to a pair of clean shorts and a toothbrush and the airplane's still too heavy, remove some of the fuel and plan an extra stop along the way. So it takes time to drain some fuel; so you'll lose more time when you stop for gasoline. Where would you rather be—safely on the ground at Halfway International, wishing you were still flying, or struggling to keep an overloaded airplane in the air and wishing you were safely on the ground?

ZERO DOESN'T ALWAYS MEAN NOTHING

Some aircraft—notably those with large, roomy cabins—are afflicted with an in-flight problem that results from all that weight pressing down right on the middle of the wing. More and more pounds could be added inside the airplane until the structural limit was reached, whereupon the wing spar would break and the airplane would proceed to the ground in at least three pieces. When the designers see this rather nasty situation shaping up, they determine the wing limit and place a restriction on the number of pounds that may be loaded inside the airplane. Thus limited—and the restriction takes the 3.8 load factor into account—there should be no problem with wings breaking off in flight.

But there's a trap here for the unwary pilot, and its jaws are sharpened by the fact that reducing the amount of fuel so that you can carry more weight in the cabin without violating gross-weight limits usually aggravates the wing-bending problem. By substituting pounds in the cabin for pounds in the tanks, you concentrate the weight pressing down on the center of the wing, and may grossly overload that rather important part of the aircraft structure.

Bearing in mind that there is a definite limit to the number of pounds a wing will support, it is possible, practical, and safe to allow higher gross weights, but only so long as the additional weight is carried somewhere other than in the cabin—such as in the wing tanks, about the only place remaining. When this is the case, the airplane's operating limitations will contain a statement such as "All weight in excess of so-and-so pounds must be usable fuel," or perhaps "Maximum zero-fuel weight for this airplane is umpty-ump pounds."*

Maximum zero-fuel weight is a term used mostly for transport-category aircraft, but it's creeping into the general aviation vocabulary as our machines get bigger. It means exactly what it says—the maximum number of pounds an airplane may weigh when there is no fuel in the tanks—and since the only other place you can put pounds is in the cabin, this restriction limits the fuselage load, and prevents the wing-breaking situation from arising. For example, an airplane with an empty weight of 2000 pounds has a maximum allowable gross weight of 4000 pounds, but a maximum zero-fuel weight of 3500 pounds. You may load any combination of fuel, passengers, and baggage so long as the total weight stays below 4000 pounds, but in no case may the cabin load be more than 1500 pounds. It's a fixed limit, based solely on the strength of the wing structure.

Local restrictions on loading will show up in many airplanes; such limits are indicated by a placard at the site, and always are included in the operating limitations. Typical are those on baggage compartments and hat shelves, where the structure can support only a certain number of pounds when the 3.8 load factor is applied. A shelf limited to 20 pounds, for example, is actually 3.8 times stronger than that, but more than 20 pounds of hats could cause structural failure if the pilot pulled out of a dive and imposed 3.8 G's on the airplane.

PLACING THE POUNDS PROPERLY

If the hand-in-glove relationship of weight and balance isn't the most important thing to be understood by a pilot, it's not far behind whatever is in first place. A seriously unbalanced load has an adverse effect on controllability, so a weight-and-balance dummy can be the world's best pilot and still come flopping out of the sky because his airplane can't possibly respond to his control inputs. Fortunately, the designers have arranged things so that you have to *work* at loading a modern airplane to a dangerously out-of-balance condition—but some pilots are very diligent. Others, the smart ones, realize that balance happens in degrees, that they have considerable control over the degree of balance, and that the closer to the optimum balance point they load the airplane, the better, faster, and more comfortably it flies.

Three items are involved in understanding, preventing, and correcting the problems of in-flight balance: center of lift, center of gravity, and the aerodynamic power of the elevator to control nose-up and nose-down movements. The elevator exerts up or down force depending on its position (streamlined or deflected into the airstream) and the speed of the air flowing across it, with a near-infinite range of elevator-control force available—from zero when there's no movement of air to one hell of a lot (even unto damaging) when the wheel or stick is moved full travel at high airspeeds. *Center of gravity* (CG) is the point at which all the weight of the airplane appears to be concentrated. *Center of lift* (CL) is the antithesis of CG, the point at which all the lift appears to be pushing upward; it moves back and forth somewhat at different angles of attack and flap settings, but for simplicity's sake, consider it as acting upward right in the middle of the wing.

In level flight at a predetermined airspeed and with the CG in an ideal position, a tailless airplane would exhibit a definite nose-down tendency, because the designers put the center of lift behind the center of gravity (you'll see why shortly). The nose-down problem is solved by adding a horizontal stabilizer, which creates just enough downlift at that predetermined airspeed to keep the nose level. This is an airspeed-of-equilibrium —everything "in trim"—so far as pitch is concerned, and with nothing changed the airplane would fly at that speed all by itself until it ran out of gas. Should the nose drop a bit, the resultant increased airspeed would move more air across the stabilizer, increase the downlift, and return the nose to where it belongs, and vice versa—the essence of pitch stability.

Given the same in-flight situation (flying at trim airspeed, CG at the same place), the pilot can create more or less downlift at the tail by operating the elevators—and the force will cause pitch change in direct proportion to the amount of elevator displacement and the distance from elevator to center of gravity. That distance is really the length of the aerodynamic lever you have available to move the nose up or down, or to prevent movement if that's your objective. At a particular airspeed, each increment of elevator displacement produces or prevents a certain amount of pitch change—nose up, for example—but don't forget that you're dealing with aerodynamic properties, and with the inevitable bleedoff of airspeed that follows when you pull the nose up, the force exerted by

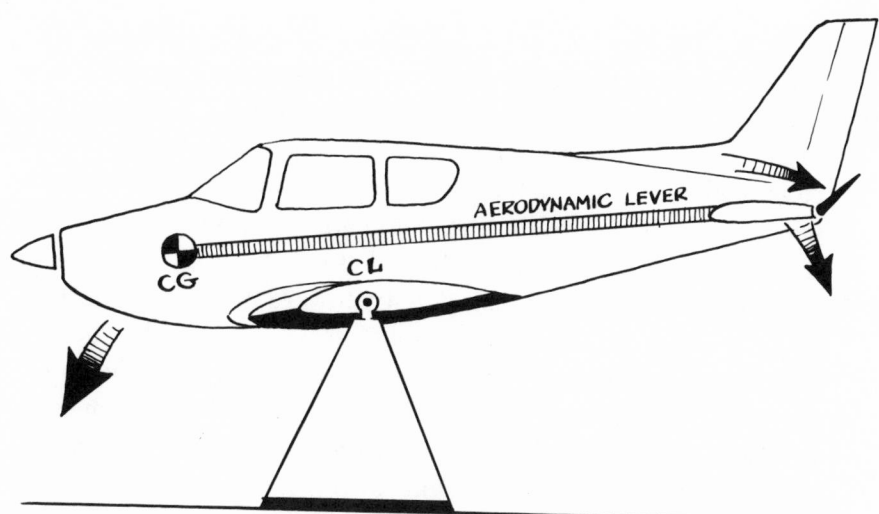

FORWARD CENTER OF GRAVITY = NOSE HEAVY

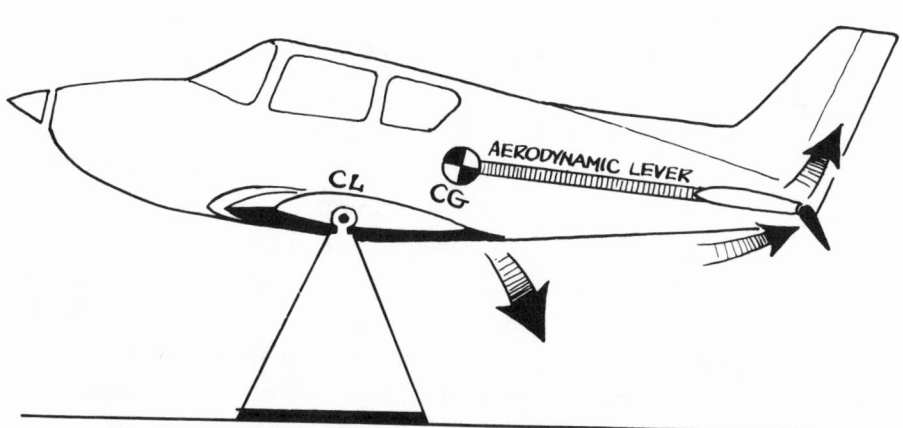

AFT CENTER OF GRAVITY = TAIL HEAVY

Figure 11

Extreme forward or aft CG locations affect both balance and controllability

---

the elevator also decreases. To maintain the status quo requires more and more elevator displacement, until the wheel is back to the stops—whereupon the built-in nose-heavy tendency takes command and pitches the nose down. That's what happens when you practice stalls; the automatic nose-down starts to take care of the problem no matter what you do.

Now move the center of gravity toward the front of the airplane—put Fat Albert in a front seat—and see what happens when you fly at the original equilibrium airspeed, the one the plane maintained all by itself. The forward movement of the CG causes the airplane to nose-down and speed up until there's enough additional downlift at the stabilizer to equalize the effect of the new center of gravity. This situation results in a new trim speed, higher than the original, which means that unless you do something, the airplane will continue to descend. But for this demonstration, the objective is to maintain the original altitude, so when Fat A changes seats, apply enough back pressure on the control column to keep the altimeter steady. The upward displacement of the elevator creates enough additional downlift to hold the nose where you want it.

Now *really* move the CG forward, and it's obvious that at some point you will run out of elevator control—with the wheel fully back, the nose is still dropping, and you can't do a thing about it except add power or speed up, or both. Carried to the extreme, this condition could get to the point where neither airspeed nor power is sufficient to keep the nose from dropping, and the airplane will descend, whether you like it or not. Consider a pilot trying to take off or land an airplane that's loaded heavily toward the nose—at the low airspeeds involved, there may not be enough elevator force to rotate the wings to a lift-producing angle of attack for takeoff, nor enough force to keep the plane from smashing onto the runway nose-wheel first at the end of the flight. A nose-heavy aircraft will usually produce a very exciting spin, and an even more thrilling high-speed recovery *if* the pilot is able to pull it out before the plane screws itself into the ground.

Forward-CG conditions are bad enough, but the things that can happen to an extremely tail-heavy airplane are nothing short of spooky. When the airplane is loaded with the center of gravity way back there, your efforts to produce or prevent pitch changes are much less effective, and the airplane becomes increasingly unstable. The aft CG location shortens the aerodynamic lever, which formerly provided crisp, positive pitch changes with only a small displacement of the control column, and when the horizontal stabilizer is called upon to change the nose attitude by itself (as in stability), it has a lot less to work with. As a result, the solid feel of the airplane is gone; when the nose starts down, you may need to pull the wheel nearly all the way into your lap to stop the drop. Then, an instant later, you're looking at a windshield full of sky. Increased airspeed helps some, but takes you into another wild world of aeronautic gyrations. With little stabilizing available from the short-levered tail surfaces, it's possible to get your full-travel control movements out of cycle with the oscillations of the airplane and, because of the higher speeds, induce load factors that could cause a stall or maybe even structural failure.

The latter catastrophe provides its own solution, of course, but the stall characteristics of an aft-CG airplane should be more thoroughly understood. When the stall occurs, the nose-down force that results from the normal positioning of center of gravity and center of lift fails to operate, or is at least weakened, because the CG has moved rearward, closer to the center of lift. The nose should drop when the wing quits flying, but there's no guarantee that it will, and when you move the stick or wheel all the way forward, the response will likely be very, very sluggish. Carried to the extreme, a critically aft CG location could result in an airplane that *cannot be unstalled.*

The same takeoff and landing problems discussed earlier

with regard to a nose-heavy airplane are still with you in the aft-CG configuration, but they're reversed. The tail-heavy pilot will discover that the normal back pressure used to rotate his airplane for takeoff will bring the nose much farther off the ground than he desires; when it's time for the roundout, the pitch movements can become so sloppy and slow that even the finest landing effort turns into a controlled crash, or worse. A tail-heavy spin will be a little less exciting—with more weight concentrated toward the rear of the airplane, the nose will remain much higher, sometimes level with the horizon in a flat spin—until you decide you've had enough and attempt a recovery. Remember that any spin, flat or otherwise, is necessarily a condition of complete aerodynamic stall, and if the airplane cannot be unstalled because of its aft center of gravity, there's trouble in River City.

One CG extreme is decidedly worse than the other; if you have a choice of too far forward or too far aft, take the former every time. Unfortunately, the much more hazardous aft-CG situation is the easier one to get into, because of the way airplanes are built. The designers cleverly place the fuel tanks in the wings, where the weight of the gasoline won't affect the balance much either way, and since at least one person must be on board to operate the airplane, the pilot's seat is located where his weight will change things very little. It's when you add other people and things that the problems begin to show up—and where can additional items be loaded in the average-type light airplane? With the exception of a few models with baggage boxes in front of the cabin, everything that follows the first two occupants into the airplane winds up seated or strapped toward the rear, and the CG moves right along with the loading. Murphy's Law at work with a vengeance: If an airplane *can* be loaded dangerously tail-heavy, it *will* be.

There are safe limits for the CG, both forward and aft, and the manufacturer must temper the load-carrying abilities of his airplane with this in mind. Enter the federal government, whose watchdog of safety, the FAA, must see these limits demonstrated before a certificate is issued for the manufacture of a particular model. With the center of gravity at the most critical locations and at varying gross weights, the airplane is put through a series of aerodynamic paces to demonstrate just how far to the front and the rear the CG can be moved and not wipe out adequate control. The flight tests usually include stalls in all configurations, spins, whatever maneuvers the pilots-to-be might get into, and a bunch more the average flyer will hopefully never experience. If, when fully loaded, the airplane is too tail-heavy to be safe, the builder has little choice but to restrict the allowable weight or its distribution (that usually hurts the sales potential) or correct the problem aerodynamically (which costs more money). In any event, when an airplane is put on the market, its papers include a document that takes into account all the possible loading situations and provides the pilot with firm guidelines for the number of pounds he can put on board and the distribution of that load to keep the center of gravity within safe limits.

IF YOU FLY FOR A CIRCUS, PUT THE FAT LADY ON A TRAIN

In general, people who are heavy enough to create a weight and/or balance problem will be too large to fit into the seats anyway, and an everyday policy of "heaviest passengers up front" will probably keep you out of serious trouble. But that's not enough in the eyes of the law, for the pilot is required to be aware of the weight and balance situation before every flight, and to take corrective measures if either is out of limits. After a few actual load calculations, you'll be able to guess body weights as accurately as the guy with the scales on the midway, and you'll also develop a deep respect for that uncomfortable feeling when your passengers show up with twice as much baggage as they said they'd bring.

All loading computations start from the same base—the empty weight and the CG location that goes along with it—and arrive at a common point of information: *total loaded weight* and a *new CG location.*

Empty weight (the airplane itself with no people, baggage, or fuel on board) and the center of gravity location in that configuration are required entries in the airplane documents, and must be revised whenever modifications or addition or removal of equipment changes either of the numbers. The CG location is defined in inches from an imaginary point close to the forwardmost part of the airplane—usually the prop spinner on a single, the nose of a twin. Since this is the point from which all balance measurements are made, it's called the *datum.*

Suppose a certain airplane weighs 1000 pounds in the empty configuration and its center of gravity is located exactly 100 inches from the datum. If all 1000 pounds could be bundled together and placed at that point, the downward force would be defined as a *moment* of 100,000 pound-inches, or the force of 1000 pounds at the end of a lever 100 inches long. It's an imaginary lever and an imaginary concentration of the weight, but by considering aircraft loads in these terms, the effect of any given weight can be determined.

From the base numbers of empty weight and CG location, it's a very simple exercise in arithmetic to add up the weights of people and baggage and fuel, *and* the effects of each one of those items in terms of moments. You need to know how much each load item weighs and where it will be placed in the airplane. For example, a 200-pound pilot would create a moment of 20,000 pound-inches if the front seat were right over the center of gravity (200 pounds acting through an imaginary lever 100 inches long). When the weights and the moments are

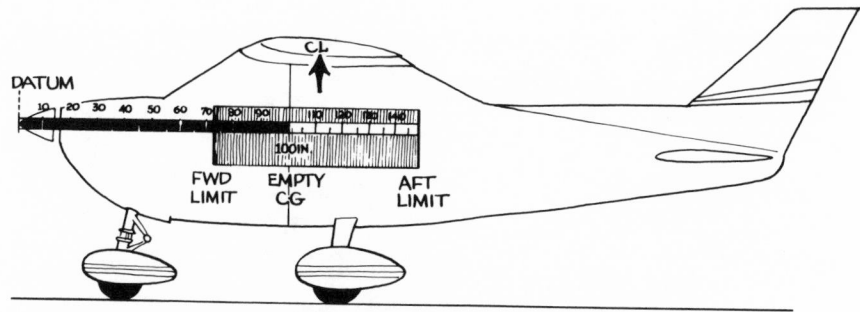

EMPTY AIRCRAFT WEIGHT 1000LBS.

Figure 12

Empty weight and the corresponding CG location are the foundations on which every aircraft loading problem begins

totaled, the new center of gravity is determined by solving this relationship:

WEIGHT times CENTER OF GRAVITY equals MOMENT

therefore

$$\text{CENTER OF GRAVITY} = \frac{\text{MOMENT}}{\text{WEIGHT}}$$

To prove the point using the previous example,

WEIGHT		times	CENTER OF GRAVITY	equals	MOMENT
Empty weight	1,000 lb ×		100 in.	=	100,000 lb-in.
Add a pilot	200 lb ×		100 in.	=	20,000 lb-in.
Loaded weight	1,200 lb ×		?	=	120,000 lb-in.
			?	=	120,000
					1,200
			CENTER OF GRAVITY	=	100 in.

With the pilot's seat at the CG, nothing changes but the weight. A passenger in the right front seat will add only weight —no shift in the center of gravity. But when you start piling

---

Aircraft Loading / 69

pounds into the rear seats and baggage compartment, the CG will move inexorably aft.

There are four methods of calculating load distribution; the first is a continuation of the problem above. Multiply the weight of each item you put on the airplane by the number of inches from the datum (the airplane builder provides a table which shows the location of each of the seats, fuel tanks, and baggage boxes) to get the total weight and total moment. Make sure the gross weight is no more than is allowed, then divide the moment (always the larger number) by the weight to arrive at the new center of gravity location in inches from datum.

Certain manufacturers provide a loading graph, which does the multiplication for you. Directly under the intersection of the weight line and a sloping line which represents distance from datum you'll find the moment for that particular item.

Since moment is the number that really tells the tale—at any weight, a larger total moment means a farther-aft CG—some aircraft are supplied with loading tables, or lists of the moments that result when certain weights are placed in certain compartments. This method frequently terminates in a table that compares total weight with a permissible range of total amounts; when the loaded CG (expressed in pound-inches) falls between the minimum and maximum moments on the table, the load is within limits.

And the fourth method is the easiest of all—no computations, no arithmetic. Just mark the weights at the proper points on a sliding plastic scale, and if the final mark falls inside the limit line on the plotter, you're ready to fly. It's so easy and nearly foolproof that maybe someday all light airplanes will come equipped with a weight-and-balance plotter.

Whether you get there by means of simple arithmetic or a plastic plotter, there's some kind of document for every airplane that provides center of gravity limits (fore and aft) at various weights. Your task is merely to check the numbers to be sure they're within those limits. If they are, go flying, secure in the knowledge that the airplane will perform safely and predictably.

But suppose that one of the straws you loaded is breaking your camel's back, or that the center of gravity is outside the limit, or both—as will usually be the case when the seats *and* the tanks *and* the baggage compartments are filled. You've no choice but to leave somebody or something at home. If the total weight is legal but the distribution is at fault, shift things around a bit until the CG is back where it belongs. (Maybe "no choice" is a bit too strong; you can go ahead with the flight, but if anything happens you are in violation of a federal regulation, probably without insurance coverage all of a sudden, and stupid. Even if an overloaded airplane staggers back to earth in one piece, the non-aviation folks on board will likely walk away vowing never to fly with you again, and you can bet they'll have some extremely unpleasant things to say—loudly, and to anyone who will listen—about little airplanes in general.)

The hazards of aircraft loading are part of the job—another piece of the responsibility you pick up when you become a pilot—but there are plenty of clues that fairly scream when a load is out of limits. For example, a sharp pilot will have a set of maximum-load numbers stored in a corner of his mind, and when the situation smacks of critical he'll go to the loading charts and make sure. If you can't see over the nose when everybody's settled in his seat, if your nose-wheel airplane has suddenly become a tail-dragger, suspect an extreme aft center of gravity and then do something about it.

A last-chance indication that all is not well in the loading department would be a very uncomfortable tendency toward over-rotation during the takeoff roll. Last chance? Yup . . . because a bad weight-and-balance problem will get worse if you take it into the air. Pros have gotten away with it, but the average ham-handed amateur pilot stands a pretty poor chance of bulldogging an overweight, out-of-balance, unstable flying machine around the pattern and putting it back on the ground safely. When in doubt, stop; and on the familiar firma of the airport terra, figure out why your airplane isn't doing things the way it should.

Propulsion Systems

4

They said it couldn't be done, but Roger Bannister did it—ran a mile in less than four minutes—and the times have been falling ever since. Even a high-school sprinter might as well hang up his spikes if he can't travel a hundred yards in ten seconds, and one of these days some super pole-vaulter will clear a twenty-foot bar. We humans have improved on the speeds at which we can move, the weights we can lift, but there's one high and shining objective that may never be reached. No matter how hard we've tried, no man has been able to fly—to lift himself from level ground, travel more than a few feet through the air, and land (crash, even) at some point no lower than where he took off.

It's a tough, maybe impossible thing to do; we're just not built for that kind of locomotion. Birds get away with it because their skeletons are ultra-light, and they have big muscles where they do the most good for flapping wings. It's a self-defeating situation—muscles big enough to overcome gravity and with enough stamina to sustain human flight would be so heavy that the super-jock involved wouldn't be able to get off the ground. "If God had meant man to fly he'd have given us wings" looks like a pretty good bet as long as you're talking about man actually *flying*, providing his own propulsion. Well, back to the drawing board.

The next best thing is some other form of energy converter to generate thrust. Enter the internal combustion engine, most likely candidate for continued use as *the* method of light-aircraft propulsion. Not very efficient, but low enough in cost and horsepower per pound, the gasoline-burning, propeller-turning reciprocating engine is going to be with us for many years.

Mechanically, aircraft engines are much like automobile engines. Okay, so they're air-cooled (because radiators and water jackets and water itself are too heavy to haul around in an airplane), and most of them are horizontally opposed engines with the cylinders lying on their sides (just as in a Volkswagen), but both auto and airplane engines perform work by changing the energy in gasoline to rotary motion; it turns the wheels of your car, and the propeller of your airplane. The dissimilarities start to show up when you consider that the aircraft engine is subject to a rapidly changing environment—what automobile is asked to operate at sea level and at 10,000 feet on the same day? The very quality of its performance (altitude capability) that makes the airplane what it is has the greatest effect on its power plant, but that effect is a negative one only when the operator doesn't understand and tries to make sea-level things happen at 5000 feet. Nor do you need fly up to 5000 feet to get your engine a mile high, performance-wise—pick a really hot, muggy day at most any airport, and watch the power output begin to sag.

Time was when a pilot needed to be part mechanic—just like early automobile drivers, who got along a whole lot better when they were able to get out and get under—but thank goodness those days are behind us. If the inner workings of aircraft engines (or any kind of engine, for that matter) are deep, dark mysteries to you, don't worry about it. After all, there are millions of automobile drivers who wouldn't recognize a crankshaft if they tripped over one, and they seem to get along just fine. What's really important is your understanding of the power-plant functions you're able to manage from the cockpit, and the application of some fundamental techniques for safe, efficient operation of the propulsion system. The engine controls sticking out of the instrument panel are a lot more than "go handles" and on/off levers; they're flight controls just as much as stick and rudder, and, properly handled, can help bridge the gap between airplane driver and pilot.

FROM STRAIGHT LINE TO CIRCLE

Unless your youth was completely misspent, you have been a reciprocating engine on more occasions than you realize—every time you've ridden a bicycle, as a matter of fact. Your knees moved straight up and down, your feet turned a crank round and round, and thrust showed up where tire and tarmac came together. The major difference between a man on a bike and an internal combustion engine is the source of energy that provides the initial straight-line motion: food for one, gasoline for the other. But the principle is the same—in an aircraft engine, *pistons* move up and down (or more likely back and forth, since a horizontal design creates a flat engine, and makes it easier for the pilot to see over the nose) to turn a crankshaft and a propeller, whose aerodynamic qualities generate thrust.

Whereas food energy is extracted smoothly and continuously as needed, a mechanical device (reciprocating engine) must operate in cycles, so an aircraft engine is designed to take in a mixture of fuel and air, compress it to raise the temperature, then ignite it to release the energy, allow the superheated mixture to expand and drive a piston for power, and exhaust the products of combustion. Those processes—*intake, compression,*

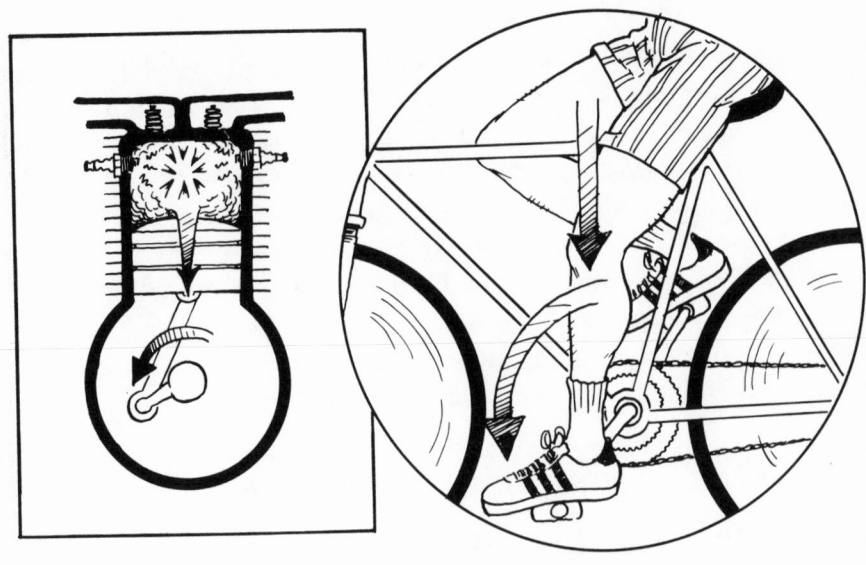

Figure 13

Principle of the reciprocating engine

power, exhaust—are the fundamentals of the four-cycle internal combustion engine. At the heart of the design is the *cylinder,* or combustion chamber, where the heating takes place. It must be strong-walled and tightly sealed so that the pressures of internal combustion can do nothing but push a piston (knee), which turns a crank . . . you know the rest of the story.

If you ever tried to ride your bike with only one foot, things got a little unbalanced, because there was no push to return the pedal to the top of its circle; in effect, you had a one-cylinder engine. With both feet working, part of each push served to bring the opposite pedal back to where another power stroke could be accomplished. A one-cylinder aircraft engine would have a similar problem, so you'll find light airplanes almost universally equipped with four-footed engines—four cylinders, two on each side of the crankshaft, their cycles arranged so that they provide pushes (power strokes) in consecutive order. And if one-after-the-other power strokes are good, overlapping pushes should be even better, so the larger engines have more cylinders—usually six, with a few eights here and there—and wind up with smoother power and more of it.

Regardless of size, aircraft engines won't run without lubrication, cooling, ignition, and carburetion. A basic understanding of these processes will help you operate the propulsion system more efficiently—and if knowing what's happening up front saves your neck someday, so much the better.

LUBRICATION AND COOLING

The oil that helps reduce friction between the moving parts must be considered the engine's blood—when there's trouble with the lubrication system, there's trouble with the entire engine. The pilot's responsibility is rather uncomplicated: Make sure there is adequate oil in the engine before starting, and shut it down any time the instruments show a loss of that precious fluid. It's the Achilles' heel of all engines—they just won't run very well or very long without oil. If the gauges show a loss of oil and you don't stop the engine, it will stop itself sooner or later. Metals expand with heat, and without oil to reduce friction, the collection of moving parts that's been keeping you in the air will swell to the point where movement stops. The engine "freezes"—locks up—and for the pilot behind it, aviation comes to a sudden, expensive halt.

The undeniable importance of proper lubrication is the reason for every aircraft checklist's reference to oil pressure immediately after the engine is started. If the oil-pressure needle doesn't come off the zero peg within thirty seconds, stop the engine and find out what's wrong. It may be only a faulty gauge (which a good mechanic can check in very little time), but if it's truly a problem in the oil system, a timely shutdown will quite likely save the engine, and perhaps your neck as well. If you ignore the no-pressure warning and the engine locks up after you're in the air, it's forced-landing time, with no power options to improve your position relative to a suitable landing spot. Complete oil-system failure doesn't happen very often these days, but complacency is as rampant as ever. Don't ever start an engine until you're certain it has enough oil to operate properly (your owner's manual specifies the absolute minimum amount), and as soon as it's running, check the oil-pressure indication . . . pretty cheap insurance when you consider the consequences.

So long as there are oil companies producing different types of lubricants, there will be controversies over which is best—multi-grade versus seasonal weights, detergent versus non-detergent, and a string of features and benefits as long as the list of manufacturers—but the answer, the most reliable answer, is as close as your owner's manual. Somewhere in those pages is a recommendation from the people who built the engine; there's no one who knows better which oil is the right one to use. Can't find the good word? A postcard to the engine factory will bring an immediate and complete response. If you're charging about the sky behind a brand-new engine, look even more carefully for the oil recommendation; using a type other than that suggested by the engine builder (or overhauler, sometimes) may void whatever warranty you bought with the engine.

The oil system is a big factor in engine cooling, in addition to its friction-reducing chore. Despite the efforts of designers and engineers, the best engines use only about 30 to 40 percent of the heat energy in gasoline; the balance of the BTUs is carried away in cooling air directed over the engine, and in engine oil passing through a heat exchanger. You've no control over this function in small engines; the oil cooler is a fixed unit, a honeycomb of small oil tubes with air passages in between. On larger installations, movable oil-cooler doors give the pilot the option of adjusting and controlling the oil temperature by regulating the amount of air flowing through the unit.

The cooling function must be considered every time you look at the dipstick. Certainly, the engine will operate safely with the minimum amount of oil in the system, but it's penny wise and pound foolish. Operating consistently on the minimum-quantity side will create problems on a hotter-than-normal day, and when you have to run the engine at high power settings (therefore high temperatures) for extended periods of time. On the other hand, a pilot who always pours in an extra quart just to be on the safe side may wind up with a higher-than-necessary oil bill and a dirty airplane; when the oil expands from heating, the extra quart may have nowhere to go but overboard through a breather line, which lubricates the belly of the airplane . . . messily. There's a recommended oil level for each engine, one that's been researched and tested, one that will suffice for all but the most rigorous types of operations. When you intend to fly in an abnormal situation (extreme heat or cold, dusty conditions, prolonged high-power settings, and the like), remember that you're asking the engine to perform outside the parameters of normal operations, and you should consult someone who knows what special protection the engine requires under those circumstances.

While oil plays an important role, most cooling is accomplished by the passage of air over and around the engine itself, specifically the cylinders, or combustion chambers. You've never really seen the cylinders in your automobile engine, because they're surrounded by a water jacket—the cooling actually takes place in the radiator—but notice that an aircraft engine has finned cylinders, to provide as much surface area as possible. The cowling serves not only to streamline things up front, but to direct cooling air into the engine compartment; in addition, most engines have baffles that channel air into the less accessible places.

If you keep the airplane moving, you'll keep the engine cool —and that's the essence of temperature control in light aircraft. (More powerful engines require more cooling air to dissipate the additional heat, so *cowl flaps* are added—air doors controlled by switches or levers in the cockpit—to enable more cooling air to pass through the engine area.) Although most installations are quite efficient and will take a lot of thermal abuse, don't expect engine temperatures to do anything but increase while the airplane is on the ground. You can help maximize the designer's efforts by holding ground operations to a minimum (especially on hot days) and always stopping the airplane with the nose pointed into the wind to let the additional airflow carry away heat.

Once airborne, the pilot's control over engine temperature depends on the sophistication of the systems. If your airplane is equipped with cowl flaps, opening (or "cracking") them a bit will admit more air and solve the problem. Should the temperature proceed toward hotter-than-you'd-like while you're climbing in an aircraft with no cowl flaps, there's little choice except to trade off some climb performance for additional airspeed and therefore more cooling airflow. It's a sacrifice, but giving up a minute or two on the way to cruise altitude is a lot cheaper than letting the engine overheat and winding up with a big repair bill. As the fates of aviation would have it, the days when cylinder-head temperatures become problems (the really hot days) are the same days when climb performance is dismally low to begin with. Ah well, the great compromise.

Cooling problems at altitude are seldom encountered; a touch of cowl flap, a slightly richer fuel-air mixture, will usually bring engine temperature down to an acceptable level. However, the cooler air of high altitudes can be misleading— especially when you fly *very* high. Even though the air temperature will be significantly lower, there is less air available for cooling. It's a problem of mass air flow, with fewer molecules to carry away heat.

In general, aircraft engines do their best work at temperatures higher than most pilots feel comfortable with. A "hot" engine will make more efficient use of the fuel because of better vaporization, and internal friction will be reduced because the lubricating oil will be somewhat less viscous, but both of these good deals have limitations. If the cylinder temperature gets to the point at which the fuel-air charge explodes instead of just burning rapidly, the resultant super-pressure can literally blow a hole in the engine or parts thereof; if the oil gets so thin that it can't keep moving metal away from moving metal, the engine will grind away its own insides—or at least scratch and scrape and score, none of which is conducive to long engine life.

A too-cool engine also has its drawbacks; less efficient combustion, and more internal friction because of unnecessarily thick oil. While these anti-benefits won't get you into trouble, they'll nail you in the pocketbook over a long period of time, because more fuel is required to deliver the same amount of power to the propeller. If you're a wintertime pilot in those parts of the country where winter is *winter*, there's not much to do except change to the proper oil when the thermometer falls, keep the cowl flaps closed and the mixture lean.

IGNITION

For reasons of safety and efficiency, airplane engines are equipped with dual ignition systems—two spark plugs in each cylinder, two completely independent sources of electrical energy. It's a safety concept, because your aircraft engine can run well (albeit less powerfully) on either system; it's an efficient system, because two sources of ignition enable engine designers to get the most energy from each charge of fuel-air mixture taken into the cylinders. On top of all that, the ignition systems are completely self-contained—so long as there are no broken parts or faulty components, you'll have sparks whenever the engine is turning.

Magnetos ("mags" in aviation jargon) make it possible. Geared to the engine and turning whenever it's turning, magnetos pass tightly wrapped coils of wire through a magnetic field and thereby generate an electric current—much the same as an electric motor in reverse. This relatively weak stream of electrons is processed by a booster coil and arrives at the spark plug with enough electrical muscle to jump the gap and ignite the fuel-air mixture. Most installations incorporate a special starting circuit to provide a hotter-than-normal shower of sparks from one set of spark plugs. More recent airplanes take care of this automatically when you engage the starter; on older models you'll often find a recommended engine-start procedure, which includes using just one magneto (always specified) until the engine is running.

Since the pilot has nothing with which to adjust or control ignition while flying, it's one of the systems that must be checked prior to every takeoff. Switch arrangements vary from manufacturer to manufacturer, but in every cockpit there's some way to isolate each ignition system so you can determine how the engine operates on that magneto, that set of plugs and wires. All mag switches have four positions: Off, Left, Right, Both. The first is self-explanatory; the other three enable selection of the left mag and its associated components, or the right system, or both systems simultaneously. Of course, Both is intended for use at all times except when checking ignition integrity during the engine runup.

At some predetermined power setting, the switch is moved from Both to Left, and the drop in engine power (manifested by a decrease in RPM) is noted. For each type of engine there is an established maximum RPM drop, which takes into account the reduced combustive efficiency of single-ignition operation. Should the engine speed decay more than the maximum, if it runs rough, backfires, or otherwise just doesn't sound right, drive 'er back to the ramp and have a mechanic take a look—ignition problems are almost never self-healing. If all is well with the left system, repeat your investigation on the right side, and be sure the switch gets put back on Both at the conclusion of the procedure—single-ignition takeoffs can be scary.

The maximum rpm drop for most aircraft engines amounts to something on the order of 5 percent, which is not a crippling power loss; consider selecting single ignition if engine trouble develops during flight. When a sick engine is gagging on a malfunctioning magneto, you might as well cut it out of the system and limp to the nearest airfield on one set of spark plugs. If the engine runs better on Left or Right, find out which mag is causing the trouble and switch it off. To prove that your airplane won't fall out of the sky on one mag, try it briefly next time you fly.

Nothing happens in a magneto-powered system until something starts to move. You're guaranteed ignition so long as the engine is turning, but by the very nature of its operation the magneto must *move* to produce a spark. That movement used to be provided by muscle power—when it came to starting-up time, the only difference between old-time airplanes and old-time cars was the shape of the crank. Well, not quite the only difference, because when something went wrong cranking an automobile, the worst to be expected was a broken arm; an airplane crank (propeller) can turn into a rotary guillotine with horrendous potential for getting your day off to a bad start.

Now for the good news: Most airplanes are equipped with self-starters to do the cranking, and storage batteries to supply the initial electrical energy. Aircraft batteries are installed for only two purposes—starting the engine and providing a supply of electricity in case the generator fails in flight. Since it's needed for only short bursts of power, an aircraft battery is not a very deep reservoir of electrical energy. With the knowledge that the starter is the biggest single drain on the battery, get to know the starting procedure well, so that the engine catches after a couple of turns every time you crank it up.

Battery energy is replenished by what amounts to an electrical pump—a generator or alternator, which also supplies power for radios, lights, autopilots, and the like. (From a functional standpoint, generators and alternators do the same job, but you should be aware of the generator's inability to produce electricity unless it is turning at a certain speed, usually around 1000 rpm. The alternator can put out adequate power at virtually any speed, but, once shut down, it is impossible to restart it unless battery current is available. Some alternator-equipped airplanes have special restart circuitry to accommodate this characteristic.) There is nothing you can do to adjust or control the output once airborne, so a preflight check of the generating system is absolutely essential. Like spark plugs, generators and alternators have seldom been known to heal themselves; the ministrations of an aircraft mechanic are strongly indicated to prevent thorough (and expensive) depletion of that rather short-lived battery.

CARBURETION

No control, no adjustment, can't do a thing about it in flight . . . If you've drawn the conclusion that engine operation is strictly an "on/off" deal for the light-airplane pilot, stand by to be corrected. There's nothing in the scheme of things propulsive that requires more pilot input, nothing over which you have more control, than carburetion—making a combustible mixture from raw fuel and air, and insuring that liquid gasoline is broken up into a fine spray en route to the combustion chambers. The people who design and build carburetors provide the parts that mix and vaporize. Enter the pilot, who must interpret the conditions at hand and operate the controls to obtain whatever performance is required. Assuming that ignition will take care of itself, the pilot has full control over the quality of the fuel-air mixture and the amount of it that reaches the cylinders for conversion to engine power.

Bernoulli's discovery of the velocity-pressure relationship that makes flight possible also made carburetion possible. In the simplest of installations, the carburetor is little more than a tube through which air is channeled to the engine. A smooth airfoil-shaped restriction (known as a *venturi tube*) in the channel creates a pressure differential that pulls fuel through a nozzle and atomizes the liquid gasoline. As the engine runs faster, it ingests more air, which increases the airflow through the carburetor, which in turn lowers the pressure even more and draws additional fuel through the nozzle. In effect, a carburetor is a self-regulated system.

An aircraft engine runs quite efficiently when each part of gasoline vapor is dispersed in about fourteen parts of air, and the carburetor approximates that ratio under normal conditions. Should the mixture become too rich (too much fuel for the air available), combustion suffers from lack of oxygen; let the mixture get extremely rich and the fire will go out, because there's not enough oxygen to support the combustion of all that gasoline. Moving in the other direction, a too-lean mixture is no good, because there's no energy in plain air. The objective is to mix in an amount of fuel vapor that, when burned, will heat the air enough to do the work at hand.

The *throttle,* by far the most frequently used engine control, is connected to a valve that opens wide for full power and closes nearly completely when you want the engine to idle. It's the "power control," the means by which the pilot regulates the

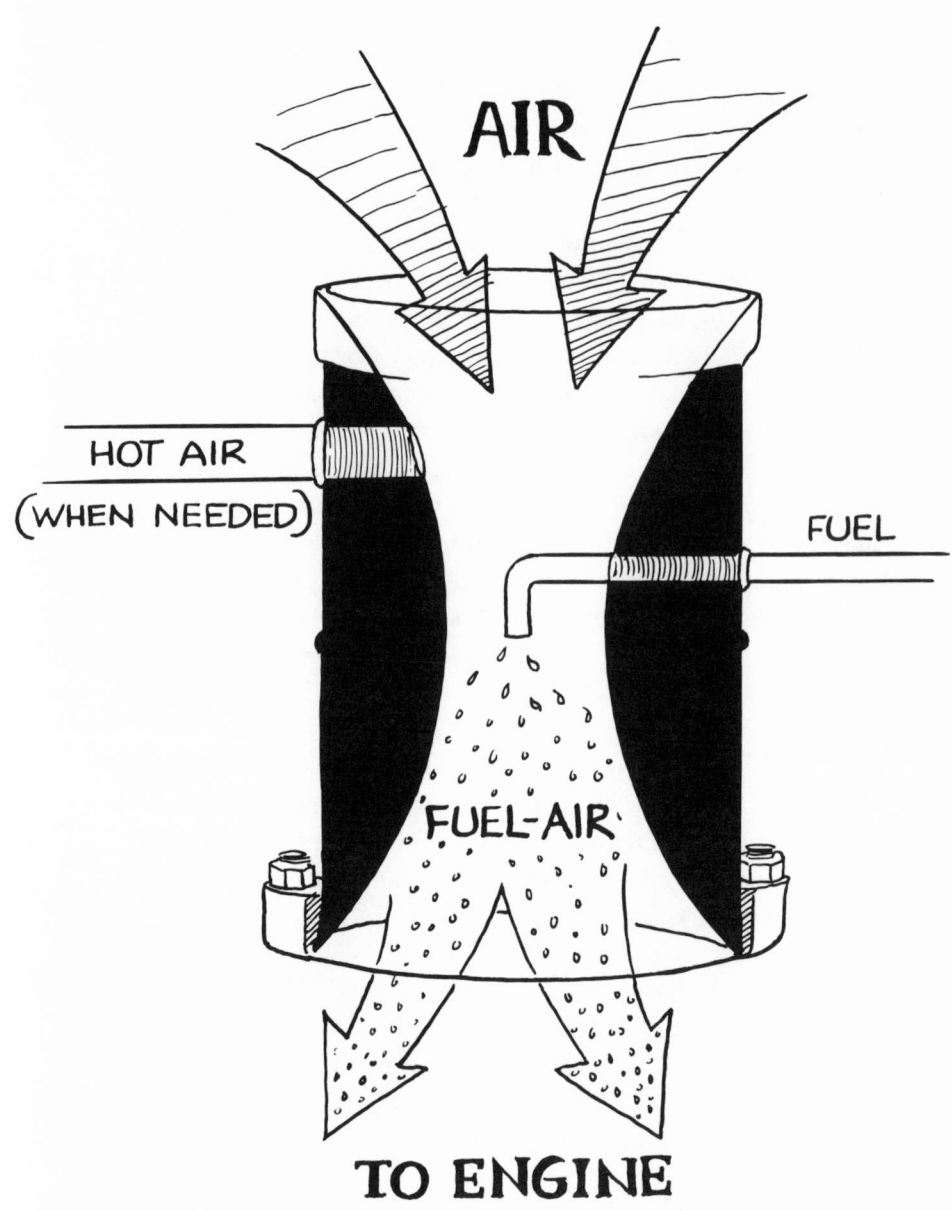

AIR

HOT AIR
(WHEN NEEDED)

FUEL

FUEL-AIR

TO ENGINE

Figure 14

The operating principle of a simple carburetor

amount of fuel-air mixture admitted to the combustion chambers. (If there's nothing else standard about power controls, you can hang your hat on this: *All* aircraft throttle linkages are arranged so that you *push to open, pull to close.*)

Quality of the fuel-air mixture is determined by a valve located in the fuel-supply line, and connected to the appropriately named *mixture control* inside the airplane. With the control in the forwardmost position (Full Rich), the fuel valve is fully opened, fuel flow is dependent on the velocity of air through the carburetor intake, and the fuel-air ratio is close to the desired 1:14. Move the mixture control all the way in the other direction and the fuel valve closes, reducing the fuel-air mixture to pure air—such an effective stopping procedure that it's the only way recommended to shut down an aircraft engine. Between Full Rich and Off are as many mixture settings as you'll ever need for an infinite variety of power requirements.

The airplane's ability to operate in a third, vertical dimension has a profound effect on engine performance. The desirable 1:14 mixture begins to change right after takeoff, and continues to change with each foot of altitude, because an increase in height is accompanied by a decrease in air density. While the weight of the air rushing past the restriction in the carburetor changes, the velocity remains essentially the same and draws a disproportionate amount of fuel from the nozzle. During a climb, the mixture gets progressively richer, until the engine literally chokes on its own fuel.

With the mixture control, the pilot can reduce the flow of gasoline and preserve the 1:14 ratio. Of course, cutting down on the amount of fuel going into the engine means the power output is also reduced. It's the price of climbing through the atmosphere and flying behind a smooth-running engine once you get to cruising altitude.

The theory is sound, but there are practical limitations. For one, the decrease in air density is very little below 3000 feet or so, and the small amount of leaning when you're flying that close to sea level probably isn't worth the effort. In the second place, oil and air aren't normally capable of carrying away all the heat of high power settings, so the engine designers rely on gasoline to provide some additional internal cooling. This is accomplished by arranging for an increase in fuel flow when the throttle is opened beyond a certain setting; the extra gasoline is not burned, but is partially vaporized inside the cylinder and exhausted, taking with it a great number of BTUs. When you move the mixture control to Full Rich, the resultant mixture is considerably richer than it needs to be for good combustion.

Most light-aircraft engines need this extra cooling whenever they're producing 75 percent or more of rated power. In consideration of the constant reduction of atmospheric pressure with altitude, decreasing air density will have claimed 25 percent of the engine's power at approximately 5000 feet above sea level. At that point it not only is safe to begin leaning the mixture, it may become necessary, to keep the engine running smoothly.

What about pilots in Denver, Albuquerque, Leadville, and a hundred other airports whose elevations put them in leaning country even before they taxi away from the ramp? They've got to adjust for it, and the way to do it is to run the engine to full throttle on the ground, lean the mixture until engine rpm reaches a peak, then enrich the fuel-air ratio to the point at which the speed once again begins to drop off. Some fuel-flow gauges are marked with the proper number of gallons or pounds per hour for maximum power at various altitudes. Which method you use is not as important as your understanding that, because of leaning to compensate for decreased air density, *the engine cannot possibly develop its full rated horsepower at any airport that lies above sea level.*

One more problem associated with the simple carburetor: When the pressure of incoming air drops as the air passes through the restriction in the carburetor throat, the temperature also decreases—as much as 30 or 40 degrees F. in some cases. If the air being drawn into the carburetor is laden with moisture and the temperature drops to freezing or below, ice is the inevitable result. It will coat the inner surfaces of the air passages, and if nothing is done the white stuff will eventually choke off the supply of air. The mixture gets richer and richer, until combustion ceases, and then you've got real trouble.

To combat the icing problem, airplanes with conventional carburetors are equipped with a carburetor heat system and a cockpit control to operate it. When this handle or lever is moved to the On position, air from a heat exchanger wrapped around the exhaust system is admitted to the carburetor throat and *voilà!*—no more ice. But you don't get off quite that easily, because the introduction of heated air means the introduction of less-dense air. Once again, you'll have to lean the mixture to retain the proper fuel-air ratio, so the ultimate price of getting rid of the ice is power. Everything costs something.

Carburetor ice is an insidious phenomenon. It is just as likely to occur on a clear day as on a cloudy one; all that's needed is sufficiently humid air and freezing temperatures inside the carburetor. At the first indication of a mysterious power loss, suspect carburetor ice and apply full heat—immediately. If the engine rpm drops off and stabilizes at a new, lower figure, there was probably no ice present—the rpm drop was due only to the introduction of hot air; but if the engine slows, runs rough for a couple of seconds, and then speeds up a bit, with the heat control still at Hot, you have no doubt burned out some carburetor ice. The rough-running period is caused by melted ice going through the engine. In a situation conducive to carburetor-ice formation, experiment to find a heat setting that will keep ice from forming, and use full heat once in a while to be certain you're keeping all of it out of the system. If the engine ever quits because it's choked with ice, you've just had your ticket punched. When exhaust heat ceases and desists, better look for a landing field close by.

THERE'S ALWAYS A BETTER WAY

Suppose man *had* been supplied with wings—there'd be an engineer or designer or inventor somewhere who would soon come up with a modification to make the system work better. The limitations of the simple venturi-type carburetor provided fertile ground for development, and today only the very smallest aircraft engines cling to the carburetor that operates on Bernoulli's principle. During the reign of the big reciprocating power plants, the pilot's mixture-leaning task was given to an automatic carburetor, which sensed the drop in air pressure as the airplane climbed and adjusted the fuel flow accordingly. Another development, intended to further circumvent complete dependence on air pressure to make things happen, was the pressure carburetor—a fuel pump supplied gasoline under pressure to the spray nozzle, and, in conjunction with automatic mixture controls, this system seemed the apex of carburetor science.

But even the most advanced carburetors feed the fuel-air mixture to an intake manifold, which distributes the combustibles to the cylinders—very inefficiently, as it turns out. From the time fuel and air come together until the mixture arrives at the cylinder farthest down the line, surface friction and condensation take their toll of the 1:14 ratio. There is no choice but to make the mixture over-rich at the carburetor to be sure that the last cylinder gets a fuel-air charge it can work with. The first cylinder is way too rich, last cylinder perhaps a little bit too lean, and *maybe* a couple of cylinders in the middle suck up the proper mixture.

When the size of the engines made it practical, and the fuel-plumbing industry make it possible, *fuel injection* was invented. A small pump runs in synchronization with the engine and sends the exact amount of gasoline to each cylinder at the proper time in the combustion cycle. There is no more wasting of fuel to be sure that the last cylinder in line gets its share; now every combustion chamber can operate at peak efficiency, and the same amount of fuel can produce more power—that's progress. It's also elimination of carburetor ice, because there's no carburetor, just an intake line to carry air to the engine—with no restrictions, no pressure- and temperature-dropping venturi tube. Of course, such a system is bound to cost more (pumps and plumbing and metering devices don't come cheap), but it's likely that the additional cost can be recovered in fuel economy and in the additional speed and/or load-carrying capabilities of an injector-powered airplane.

It may be distasteful philosophically, but wars have always spurred technology, and aviation technology has been no exception. A good example is the high-altitude performance that became an absolute necessity for World War II aircraft, and that spawned the *supercharger*. All air-breathing engines have altitude limitations—as you've seen, when you begin to run out of pressure you begin to run out of power—but the problem can be partially solved by supplying higher-pressure air to the intake system, and originally this was accomplished by large centrifugal "air pumps" deep inside the big military engines and driven by engine power. Today's superchargers are almost exclusively *turbosuperchargers*, turbine wheels driven by high-energy exhaust gases and connected to rotary air compressors. There are superchargers that can be switched on and off as aircraft performance needs change; others operate all the time, with safety valves to keep air pressure within specified limits.

The turbocharged aircraft offers its pilots some pleasant

options. Since it makes sea-level air pressure available at nearly any altitude, he doesn't suffer the loss of power that hobbles the non-supercharged fleet; he can obtain sea-level engine performance on takeoff from even the highest airports in the land (careful, wings and propellers aren't supercharged—they'll still perform in accordance with actual altitude); he can cruise at considerably higher speeds; and, maybe most important, he can climb above a great deal of the nasty weather that exists in the lower layers of the atmosphere.

There's no such thing as a free lunch, and you can expect to pay considerably more at the outset. You'll also need a personal oxygen system on board to keep the people healthy at turbo altitudes, and superchargers use more fuel in the long run. Of course, you'll have *traveled* a heck of a lot farther in that long run; dollars-and-cents considerations will be different for each pilot.

GENTLEMEN, START YOUR ENGINES!

Making an aircraft engine come alive involves nothing more than supplying the three necessities for combustion—fuel, air, and ignition—in the proper quantities at the right time. Sparks are always available (unless the pilot forgets to turn on the switches—and don't think it can't happen to you!), so your task is to set the throttle and mixture controls in hopes that a combustible fuel-air ratio will get to the cylinders during the first couple of engine cycles. "In hopes" is used advisedly, because every start will be different. The engine fires will light on a range of mixtures, but the smooth, quick starts that save batteries and demonstrate a pilot's know-how are the ones that result from setting the controls correctly the first time.

With the mixture control at Full Rich, the very first turn of the engine begins to draw fuel through the carburetor. But on a cold start (the first start of the day, or after the engine has been shut down for more than an hour or so), the intake system is probably not warm enough to vaporize the gasoline fully, and it may take a lot of cranking (and battery power) to get adequate fuel vapor into the cylinders. A combination of cool engine, cool gasoline, and cool air may inhibit vaporization to the point where it's impossible to produce a combustible mixture from the gasoline being drawn through the carburetor. In this situation (which is encountered more often than not), the engine must be *primed.*

All simple-carburetor engines are equipped with a priming pump in the cockpit to inject raw gasoline directly into the intake passages—a bit like wafting ammonia under a fainter's nose, it provides enough fuel vapor to get the engine's attention. After a cycle or two on the prime fuel, the engine can usually continue running on the normal fuel supply. The primer has no gauges or meters; "enough" is determined by the number of strokes of the pump, and after a few primed starts you'll have a good idea of how much raw fuel your engine needs. The objective is to provide just enough to get things going the first time, and temperature is the key—cold air, cold engine, cold fuel will obviously require more prime than a situation in which any of the three elements is warm. On a cold day it's difficult to over-prime, but be careful—with that much raw fuel inside the engine (and probably on the ground as well, because the excess will puddle under the engine), the potential for fire is extremely high. On a very cold day, one cold enough to freeze the ears off a brass monkey, you may have to keep the engine running with occasional shots of prime until things warm up enough to promote vaporization.

Automobile engines need a shot in the arm too—that's why the starting instructions suggest you push the accelerator pedal a couple of times before turning the key. Connected to the pedal is a small plunger called an *acceleration pump,* after its primary function of providing an extra surge of fuel when you come down hard on the pedal to go faster suddenly. Aircraft engines have a similar arrangement, and it makes a handy primer for those times when just a little extra fuel is needed for starting. On warm days, or when the engine has been shut down for only a short while, pump the throttle briskly two or three times just before you engage the starter.

Assuming that everything is set up properly—mixture at Full Rich, ignition switches on, throttle positioned for 1000 rpm— the engine should start after just a couple of turns. If nothing happens within ten seconds, you might as well stop cranking— there's something wrong with the fuel-air mixture going to the cylinders. On a cold day, it's likely you need more prime; give the primer another shot or two and try again. On the other hand, you may have over-primed and the engine is choking on the super-rich mixture. Either wait a short while to give the extra gasoline a chance to evaporate, or crank again with the throttle halfway open—the same procedure you've probably used in an automobile with flooded carburetor. *Caution*: A half-open throttle represents a lot of power; be ready to retard it just as soon as the engine catches.

FUEL-INJECTED ENGINES

No-carburetor pilots are faced with the same basic problem at engine-start time—provide the proper combination of ignition, fuel, and air—but the technique for starting a fuel-injected engine is considerably different. While ignition and throttle functions are unchanged, the mixture control becomes a direct fuel-metering device.

With the throttle positioned for 1000 rpm and the boost pump on, prime an injected engine by moving the mixture control to Full Rich for a few seconds (until flow registers on the fuel-flow gauge), then back to Off. Engage the starter, and as soon as the engine fires, return the mixture to Rich. It doesn't have to be done in a panic, but rapidly enough to bridge the combustion gap between prime fuel and the normal supply. Again, a little experimenting will help you decide how much fuel is enough—a cold engine requires more prime on a cold day.

The dirtiest words in the fuel-injection vocabulary are *vapor lock:* tiny bubbles of gas vapor trapped in the fuel-supply lines, where they block the path of cylinder-bound gasoline. The problem is usually encountered on a hot day when the engine is being restarted after a short rest. The fuel distributor and the supply lines are almost always located on top of the engine, where heat can boil the fuel to form those troublesome bubbles. When a true vapor lock exists (usually indicated by lengthy cranking with no combustion), there's little to do except open the valve (mixture control), turn on the fuel-boost pump to force the bubbles out of the lines, and start all over again.

An injected engine that has been shut down for only a few minutes (maybe up to thirty, depending on the air temperature) is a red-hot prospect for *overloading* (too much fuel). Bear in mind that a hot engine turns liquid gasoline into vapor instantly, and that whatever fuel remains in the lines is probably sufficient to accomplish an engine start. A hot engine is therefore already primed, so make your first starting attempt with the mixture control off. Chances are very good you'll fool the engine—and as soon as it catches, move the mixture to Rich.

When you anticipate a short turnaround (very little time between stop and restart), shut the engines down by setting the throttles at 1000 rpm and pull the mixture to Off, turn off the ignition switches, and go about your business. When you're ready to start again, don't touch a thing (except ignition switches)—throttles are set, mixture still at Off—and the engine should fire immediately, whereupon you must advance the mixture control to keep it running. If this procedure doesn't work, go back to the original starting sequence.

WAY UP NORTH

Airplane engines get progressively harder to start as the temperature drops. Fuel vaporization and battery power go down with the thermometer, while oil approaches the viscosity and immobility of glue. Remedies for wintertime woes at the airport can be listed in descending order of impact on the pilot's pocketbook: Keep the airplane in a heated hangar, preheat the engine with a mobile unit, use an auxiliary power unit (battery cart), or take the oil and battery home with you—that's what the bush pilots do.

PROPELLERS

After the inventive minds of early aeronautical engineers had considered and tried every kind of thrust producer from flapping wings to huge oars, they settled on the propeller to convert engine power to thrust. Airfoils that generate lift in a horizontal direction, propellers operate on the same principles and have the same limitations as any other airfoil. Rotate the blades faster, increase the surface with larger blades or more of them, increase the angle of attack by twisting the blades for a bigger bite of air, and they'll produce more thrust—up to a certain point. Just like a wing, every propeller has its critical angle of attack, and exceeding it results in a stall. Propellers don't quit flying the way a wing does, but there are times when the blades are doing little more than stirring up the air.

Light training aircraft are equipped with the simplest propeller imaginable: a solid block of aluminum sculpted to an airfoil shape, with absolutely no moving parts. With the blades' angle of incidence therefore unchangeable, such a prop represents a compromise of power and speed. The angle is low enough to let the engine build up to its maximum speed for takeoff (all aircraft engines get stronger as they run faster) and high enough to produce a reasonable airspeed at level-off, when the high-power demands of climbing have been satisfied. It's something like driving a three-speed car that is stuck in second gear: You give up the power of first gear and the top speed of high gear in exchange for *not* laying out the dollars to get the transmission repaired. With airplanes, the cost and weight of a changeable-pitch propeller (an aerodynamic gearshift, if you will) can be justified only on the heavier, more expensive models.

A fixed-pitch propeller is one with the engine. It's bolted to the drive shaft and is directly responsive to power changes, speeding up when the throttle is opened, slowing down when it's closed. With the same throttle setting, the airplane-engine-propeller combination will go faster when diving and lose speed in a climb, in direct variance with the thrust load placed on the entire propulsion system. The amount of thrust is indicated by engine speed on the cockpit tachometer; in general, more rpm means more thrust. Whether that thrust is used to increase speed or climb rate is up to the pilot.

CONSTANT-SPEED PROPELLERS

The next step in propeller development was a good deal for the pilot who needed a lot of power on one flight, a lot of speed on the next. Known as a *variable-pitch prop,* it featured twistable blades, which gave the pilot a choice: a low-pitch, high-rpm, high-power angle for lifting heavy loads (paid for with reduced airspeed at cruise) or a high-pitch, low-rpm setting, which made for longer takeoff runs but got the pilot home in time for dinner. At first strictly ground-adjustable, the one-or-the-other propeller control was later moved into the cockpit so that in-flight changes could be accomplished.

Today, all propeller-driven twins and the higher-performance singles are equipped with *constant-speed props.* The pilot controls a governor system that maintains the desired rotational speed by automatically varying the angle of the propeller blades. Within reasonable limits of airspeed and power, the governor holds a selected rpm indefinitely; you never notice the minute changes in propeller pitch as the system adjusts to constant variations in air loads.

Since the propeller rpm stays the same regardless of engine power, an additional instrument is required for a constant-speed installation. The *manifold-pressure gauge,* calibrated in inches of mercury, measures air pressure in the intake manifold and is a direct indicator of throttle position. Wide open, the throttle admits full atmospheric pressure to the engine and the gauge reads approximately 28–29 inches on a standard day at sea level (there's some pressure loss due to friction and bends in the intake plumbing). The throttle can be adjusted to any setting the pilot wants, and the manifold pressure is correlated with the rpm to determine the power being produced. There is an infinite number of power settings available.

For takeoff, the prop control (another lever or push-pull knob on the panel) is positioned for high rpm, and when the throttle is opened the governor allows the propeller blades to twist until they're absorbing all the power the engine can produce. As the airplane starts to move, the task of overcoming inertia is reduced, and the propeller tends to speed up in response to the lightened load. But the governor is extremely sensitive to speed changes, and increases the pitch of the prop blades just enough to nullify the slight overspeed. Now the engine is working just as hard, but for each rotation of the propeller more air is being thrust backward due to the increased angle of attack. This process continues automatically until the airplane is stabilized at whatever airspeed you desire for climb—the propulsion system actually gets more efficient all by itself as the airspeed increases. (Carried to the extreme, there's some airspeed at which the prop blades reach their critical angle of attack for maximum engine power. When you get to that point, the prop can't provide any more thrust, and you've flown your airplane as fast as it will go. That's one of the factors that limits the world record to 482 miles per hour in a big-engined but propeller-limited fighter airplane.)

Climb power—usually about 75 percent of maximum—is set by first retarding the throttle to a specified manifold-pressure reading. Once again, the governor senses a change in loading and flattens the blade angle to maintain the preset rpm. The second part of the power-change procedure consists of slowly moving the prop control back and resetting the governor for a new, lower rpm, which it will again maintain, being an absolute slave to the law of physics.

It's the best you could want: plenty of power for takeoff, and abundant speed at cruise.

On those increasingly rare occasions when an aircraft engine fails (or is purposely shut down by the pilot because of an impending failure), the propeller becomes a metal windmill. Instead of engine power producing thrust, the process is reversed: Wind power (airspeed) turns the engine, now reduced to nothing more than an air compressor. No useful work is being done, and the airplane will either decelerate or descend (sometimes both) in accordance with the large amount of drag from the powerless prop. This is not much of a problem on single-engine airplanes—with only one engine to start with, a power failure makes you an instant glider pilot—but a windmilling propeller is intolerable for a twin, so the *full-feathering prop* was developed. When the feather feature is selected by the pilot, hydraulic or electric forces twist the blades 90 degrees to streamline with the airflow, stopping the engine and eliminating the drag.

The *reversible-pitch propeller* is another development in propulsive technology, a clever arrangement whereby the blades can be twisted (again, with hydraulics or electric motors) beyond the zero angle of attack to generate reverse thrust. With such a system, engine power can actually be used to help stop the airplane once it's firmly on the ground. Not as efficient as forward thrust because of the flat back side of the prop blades, reverse thrust will nonetheless decelerate the airplane rapidly, saving tires and brakes in the process. Reversible propellers are just now filtering down to the general aviation fleet, and it's likely you'll see more and more little airplanes backing into their parking spaces.

EFFECTS ON AIRCRAFT

If a propeller could somehow produce thrust without rotary motion, the pilot's lot would be quite different. Especially during those regimes of flight when great inertia loads must be overcome, the resultants and counterforces of thrust become very real and apparent. The pilot who doesn't understand the origin of these forces or how to deal with them soon finds himself being taken for an airplane ride—and not necessarily in the direction or manner he intended.

Disregarding for the moment the weathervane effect caused by crosswinds, the nose of a propeller-driven airplane will try to swing to the left on every takeoff, because of the whirling airscrew up front. To begin with, the rotation of the blades (clockwise, as seen from the cockpit) imparts a corkscrew motion to the airflow. At low speeds, the fuselage is completely enveloped in a *spiraling slipstream*, which always strikes the vertical stabilizer and rudder on the left side, and the unbalanced force shows up as a strong tendency for the nose to yaw left.

The second force of concern here deals with the fact that the propeller blades, no matter how many or whether they are fixed-pitch or constant speed, cleave the air at identical angles —the right-hand blade must always describe the same angle in reference to the prop hub as the left-hand blade. When the airplane is moving down the runway on all three wheels and the propeller is rotating perpendicular to the ground, the ascending blade (on the pilot's left-hand side) has the same angle of attack as the descending blade (on the pilot's right-hand side). In this configuration, you can consider the propeller a disc with uniform thrust production all the way around.

When you raise the nose a bit to fly off the ground, there is a short period of time when the entire airplane is moving parallel to the runway but the propeller disc is rotating at some angle other than perpendicular. The angle at which the prop blades are mounted can't change, so during the transition period the descending blade (right side) experiences a higher angle of attack than its counterpart on the other side of the disc. All other factors being equal, a higher angle of attack produces more lift (in this case, thrust) and increased pulling power on the right side of the prop disc. Net result, the nose swings to the left. This is even more pronounced when you're taking off in a tail-wheel airplane, because the prop disc is producing *asymmetric thrust* from the very first revolution.

Even at cruise airspeeds, most aircraft fly in a slightly nose-high attitude, so designers included a feature to help offset this near-constant tendency to yaw left: They angled the vertical stabilizer just enough to counteract the force of asymmetric thrust at the normal cruise speed. Satisfy your curiosity by noticing the airspeed at which your airplane will fly straight ahead with no rudder pressure. When you slow down, the nose will swing to the left—the offset isn't effective enough—and an airspeed above the magic number will always cause the nose to swing to the right. That's one reason for rudder trim on all but the lowest-performance trainers.

Lesson to be learned: At low airspeeds and high power settings, a propeller-driven airplane *will try to yaw to the left,* no matter what its attitude at the time; whenever the airspeed goes above that for which the vertical stabilizer is offset, *the nose will swing to the right.*

There are other forces at work—gyroscopic effects from the propeller, and torque (which produces a slight rolling tendency that is effectively countered by different angles of incidence at the wingtips)—but the spiraling slipstream and asymmetric thrust account for the greater part of your flying machine's tendency to head for the left side of the runway on takeoff. How strong is that tendency? That depends on how rapidly you open the throttle, how much power is available, the vigor of your pitch change during rotation, and the weathervane effect of the wind. How much will the airplane turn? Not at all, if you understand what's happening and apply whatever rudder pressure is needed to keep the nose on a point. Expect the same effect when you're slow-flying or practicing stalls. If it takes a footful of right rudder to make the nose stay put, *make it happen.*

OTHER PROPULSION SYSTEMS

There's a better way to move an airplane, a way to eliminate heavy combustion chambers, cooling problems, spark plugs, vibration, and those troublesome propeller blades. That way, of course, is the jet engine—a propulsion system that provides thrust with only a handful of moving parts, and with none of the annoying (and sometimes dangerous) characteristics of propellers.

It's no trick to simplify the explanation of a modern jet engine. Air is taken in at the front of a tube, heated by compression, mixed with fuel and burned, and exits the engine going faster than when it came in. The action of the rearward-rushing air is balanced by an equal and opposite reaction against the inside of the engine, and thrust occurs. On the way out, the exhaust gases pass through a turbine wheel to power the compressor (that's why they're properly called *turbojets*). Once the fire is lighted, there's no need for ignition; the combustion process is continuous and self-sustained.

A jet pilot has no controls to manipulate the fuel-air ratio—mixture control is in the hands of an automatic fuel-control unit—and he adjusts power with a simple thrust lever. Engine output is indicated on a gauge which reads in percent rpm, or in some cases on an EPR (Engine Pressure Ratio) gauge, which shows the pressure of air leaving the engine as it relates to air pressure at the inlet. Jets are rated in pounds of thrust instead of in horsepower.

Like other air-breathing engines, the turbojet has altitude limits, but it gets more efficient in the super-cold air of the higher levels, to say nothing of its generally increased performance as airspeed builds up and the ram effect of high speed adds to the compression inside the engine. In general, jet power is most applicable and effective where speed is important.

First-generation jets were little more than sophisticated blowtorches that relied solely on the acceleration of air through the engine. They were also noisy, literally ripping the air with the near-sonic thunder of their exhaust. Bigger "straight-through" engines produced more thrust and even more noise, so the industry came up with a family of jets that propelled larger volumes of cooler air to the rear at lower velocities—the *bypass* or *fan* engines. Much more tolerable to the ears, these jets paid the piper in terms of top speed (still, 500 knots or so ain't all bad!), but they were capable of pro-

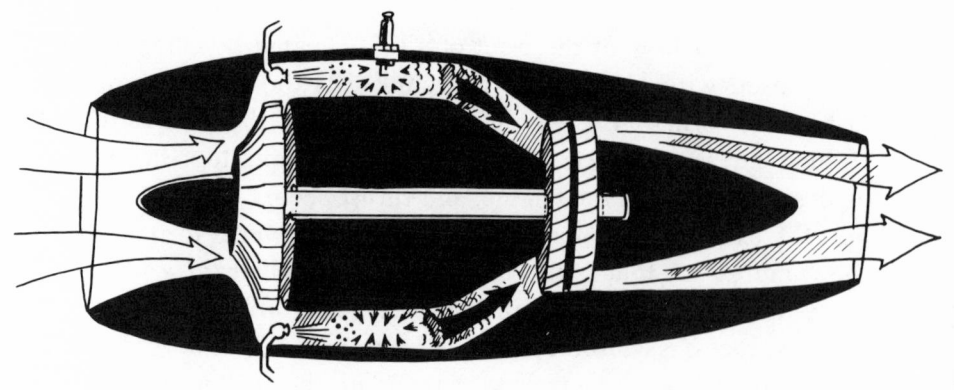

Figure 15

The turbojet engine—simplicity itself

ducing unprecedented amounts of thrust in consideration of engine weight and fuel consumed.

Between the relatively low-speed propeller system and the high-speed turbojet is an economic/performance gap bridged today by a hybrid—the turboprop (or "jet-prop," as the airlines' PR folks prefer to call it). Simply a jet engine driving a propeller, a turboprop combines jet reliability, simplicity, and lightweight power with the instant acceleration and low-speed performance of a reciprocating-engine-powered airplane.

But the gap is closing—the blades that move larger quantities of cooler, lower-velocity air through fan-jet engines have begun to look more and more like multi-blade fixed-pitch propellers. The fans on today's jumbo jets provide a surprisingly large portion of the thrust—75 percent or more in some cases—that enables these aluminum monsters to operate from not-very-long runways. The turbojet portion of the engine is no longer the roaring proponent of raw thrust, but a rather ordinary power producer buried deep in pinwheels of aluminum-alloy blades.

The pilot who has just paid a lot of dollars to have his reciprocating engine overhauled or replaced after only 1200 hours of use hopes fervently that jet engines will soon show up in light aircraft. Those hopes are based on jet-engine lives measured in thousands of flight hours. But the initial cost of a turbojet propulsion system is staggering, and can't be justified for low-speed aircraft. Unless we completely exhaust our sources of petroleum, the reciprocating gasoline engine has a firm hold on aviation's future.

Part Two

POINTS OF ORIENTATION

The Indispensable Maneuvers:
Takeoffs and Landings

Flight Instruments
Flying When You Can't See Outside
Weather
Aerial Navigation

The Indispensable Maneuvers: Takeoffs and Landings

5

There's no way to avoid takeoffs and landings—at least one of each on every flight—and there's probably no other pair of aeronautical exercises that so mirrors and illuminates a pilot's skill . . . or lack thereof. Takeoff can be a sudden separation of airplane from runway, or a smooth change from rolling wheels to soaring wings. A landing can be little more than a controlled crash, or it can be so gentle that your passengers will have to ask, "Are we on the ground yet?" The pilot who really understands what's happening during takeoff and landing becomes a manager of aerodynamic resources, controlling and directing them, blending them with the forces of nature and the laws of physics. This sounds like a big order, and it is. A lot can go wrong, so don't count on being consistently good at takeoffs and landings for a long time. Smooth, safe, purposeful transitions from ground to air and back again are works of art, created at the expense of experience and practice.

True, there are some numbers involved—every airplane has its own minimum liftoff speed, its own stalling speed—but there's also a lot of technique, the little touches that the old-timers maintain you can feel only in the seat of your pants. And they're right, except that they didn't go far enough: The sensory inputs that make for good takeoffs and landings come from eyes, ears, control pressures—everything the pilot can bring to bear on the task at hand.

Then there's this business of a "normal" takeoff or landing. Pilots, particularly instructors, are very fond of the term ("Okay, let's do a couple of normal landings"), but your definition of that maneuver is just as valid as the next guy's, although the results may vary considerably. For the purpose of this discussion, a *normal takeoff* is one in which the airplane is allowed to fly off the ground on its own terms—that is, when the wing is ready to lift the number of pounds on board. And when a gliding approach terminates in a power-off stall an inch or two above the runway, that's a *normal landing*.

Once the normals are mastered (or, more likely, gotten familiar with), you've a base of experience and some measure of skill that can serve as a springboard to the more demanding exercises: maximum-performance takeoffs and landings. Here the pilot explores the furthest corners of his aircraft's capabilities (often his, too) and puts that knowledge to work to produce the shortest possible takeoff roll, the steepest possible climb-out, and the shortest possible landing roll. You're into the calculated-risk areas of aircraft performance now, because you'll be demanding the utmost of the machine; sloppy handling and errors in technique and judgment can be disastrous. Good instruction, plenty of practice, and thorough understanding are the keys to successful operations in less-than-ideal airport conditions.

THE NORMAL TAKEOFF

The introduction of the nose wheel (really a reintroduction, since all the early airplanes were equipped with tri- or quadricycle landing gear) was somewhat unfortunate in terms of the "drive it off, drive it on" concept that it engendered in the private sector of the aviation community. With the little wheel up front, the pilot can maintain a familiar, comfortable level attitude until an airspeed is reached that guarantees flight, then rotate the nose of the airplane upward, quite literally pulling it off the runway. That kind of takeoff falls very short of demonstrating the fine art that flying can and should be. A well-done normal takeoff is a gradual increase in lift until the ship rises smoothly from the runway. When liftoff comes as a pleasant surprise to the pilot, that's *flying*.

So far as the wing is concerned, takeoff is the process of increasing the angle of attack and the airspeed until the lift produced is just a bit more than the weight of the airplane. Some aircraft sit on their wheels with the wings angled upward enough to provide the necessary lift with no further increase in pitch attitude; you'll also notice this situation when an airplane is heavily loaded, the more-aft-than-usual CG location holding the nose higher, so that, with enough airspeed, the machine will fly. But, more than likely, the pilot will need to increase the pitch attitude somewhere in the takeoff roll.

Early in the game, you should always line up on the center line of the runway and get all your ducks in a row before starting the takeoff. If nothing else, this will slow down the occurrence of events. You might as well start this exercise with the pilot ahead of the airplane.

With the engine at a comfortable idle, it won't take much if any brake pressure to keep the airplane from creeping forward, but as soon as you decide to take the plunge, consciously take your feet off the brake portion of the rudder pedals and plant your heels on the cabin floor—any situation that requires braking action to keep the nose of the airplane going straight down the runway is a situation untenable for a beginner. Full power should be applied for every takeoff—not slowly, but smoothly, with the throttle opening just as rapidly as the engine will take it without stumbling—and immediately you will notice the need for directional control: When there's no wind at all, or when it's blowing right on the nose, propeller-driven airplanes will swing their noses to the left. With the airstream from the prop washing over and around the rudder, you've plenty of control with the pedals, so apply right rudder in whatever amount is needed to keep the nose aimed right down that center line. With experience, two things will happen with regard to directional control on takeoff: First, you'll begin to anticipate the need for right rudder and the swing won't come as a surprise; second, you will use peripheral vision to pick up clues in addition to the white stripe down the middle of the runway, and the excursions in yaw from that point on will diminish dramatically.

Separating for just a moment the three dimensions of aircraft movement, consider the effect of increased airspeed on a given amount of rudder deflection. You've done a good job of noticing and correcting for yaw as power is applied, but if you maintain that rudder pedal position and allow airspeed to build up, the aerodynamic force exerted by the rudder will increase, and the initial correction will rapidly become a gross overcorrection. All other factors staying the same, you will have to back off more and more to keep the nose straight—which is the name of the game anyway. So the answer to "How much rudder pressure should I use on takeoff?" is simply "Whatever is needed to keep the nose on a point"—and you'll have to stay sharp to make that happen, because a lot of things are changing at the same time.

(By the way, take note that you're concerned with "straight ahead" in terms of reference points straight ahead of *you*, not the nose of the airplane. In most trainers you'll be sitting on the left side, and attempts to keep the airplane's nose lined up with the center line will invariably result in a turn to the left. Pick a line of rivets, or maybe the cowling hinge, to provide an aiming reference. If there's nothing out there to use, work with the rudder pedals until the runway center line appears to be sliding backward between your knees. This will usually take care of the offset and help you maintain a straight line.)

Next turn your attention to the wings. The aerodynamic forces at work during the takeoff roll are compromising, constantly changing, and so subtle in their interaction that few pilots can sense the fine tuning that would produce optimum performance. Ideally, the nose should be rotated upward a small amount just as soon as forward speed is achieved; the wings would begin generating lift right away, but the induced drag would hamper the acceleration. The increase in angle of attack that represents an even trade-off between drag rise and production of lift is probably impossible to achieve with human effort, so you should settle for something less. If you're smooth, if you practice constantly and really develop a feel for your airplane, the "something" won't be very much less.

Just as soon as you can feel the elevators come alive, just as soon as back pressure on wheel or stick causes a change in pitch attitude, start the nose upward. The amount will vary from one plane to the next, but for most light single-engine training aircraft it's sufficient to bring the top of the instrument panel up to the horizon and hold it there—not all at once, because if you put it there prematurely, the drag buildup will work against you. After a few tries, noticing the airspeed at which your airplane leaves the ground in this attitude, you'll be able to develop that smooth, unbroken increase in pitch that will bring the wings to the proper angle of attack at the same time that the airspeed arrives at the right number.

A fixed elevator displacement as the airspeed increases will give you the same problems as holding the rudder pedals in one position: The wheel pressure that brings the nose up to the horizon will keep it right on going if you don't do something, such as relaxing some of that pressure. Press forward if you have to, so that the nose stays right where you want it to be. With experience, you'll discover an elevator-trim setting that will average out the pressures on takeoff, but since a change in center-of-gravity location will change both trim setting and wheel pressures, don't count on their being the same for every departure. No matter what, do whatever you have to do with whatever you have available (flight controls) to make the airplane do what you want it to do. The first few times you take charge of both directional control and pitch changes, you may feel you're trying to rub your belly with one hand and pat your head with the other (What crazy kind of automobile would turn left every time you step on the gas?), and in a sense you are, but that's what three-dimensional transportation is all about. With practice, you'll find that doing two things at once isn't so tough—after all, you can chew gum and walk, can't you? And, in your spare time, keep the wings level, too.

Sooner or later, liftoff will happen all by itself, the inevitable result of placing the wings at the correct angle of attack and allowing the airspeed to increase to the point at which it generates more lift than weight. Liftoff will be smooth and sure if you let it happen when the wing is ready. It will happen sooner with a light airplane or a strong headwind, later when you've a heavy load or no wind at all. You can pull an airplane off the ground before it's really ready to fly—the insidious phenomenon of ground effect is a trap built into every takeoff—but it probably won't stay airborne very long, or very controllably. You can also hold the nose down with wheel pressure until the airspeed is well above that required for flight, then yank back on the control column and leap into the air like a goosed gazelle—spectacular, but also tough on nose wheels, which aren't built for such high speeds, and non-pilot passengers, who didn't bargain for that much of a thrill. The hold-it-down-and-pull-it-off procedure also convinces those who know better that you're a driver, not a pilot.

The pitch attitude that broke the shackles of gravity will continue to do good work for you in the air; while it won't produce a record-breaking initial climb (remember, this is a *normal* takeoff procedure), it will allow the aircraft to accelerate rapidly through the airspeeds required for liftoff, then best angle of climb, best rate of climb, and finally to your normal climb airspeed if it's higher than best rate. You have the option of stopping the acceleration at any point along the way when that airspeed is the best one for a particular situation.

In an aircraft with retractable landing gear, pick up the rubber rollers just as soon as you're comfortably airborne. This is a judgment call, and the amount of runway remaining probably has more to do with it than anything else. With thousands of feet of concrete ahead at liftoff, leaving the wheels down a little longer may ease the pain of an immediate return to the runway in case of engine failure right after liftoff. But most of the runways you'll be using aren't that long, and since you paid a lot of money for wheels that go up and down, pull them up and enjoy the extra performance.

A normal takeoff isn't complete until the airplane is well established in its climbing configuration. For simple airplanes —fixed landing gear, no flaps, fixed-pitch propeller—that means a pitch attitude to produce the normal climb airspeed at full throttle. Although the complex aircraft are certified to run all day with the engine at takeoff power, it's just not good sense or good operating practice to climb out with the power handles up against the stops. With a controllable-pitch propeller system, reduce power to the recommended climb setting as soon as the landing gear is retracted or when climb speed is established in a fixed-gear airplane. In days of old when men were bold and engines weren't predictable, this wasn't a good idea, but today there's no reason to "floor it" for long periods of time after takeoff. Besides, the airport's neighbors will deeply appreciate the reduction in noise level.

So long as the wind is blowing right down the runway, the only difference you'll notice on takeoff into a strong wind is that the aircraft's performance improves in every department except groundspeed. Everything will happen faster in direct relationship to the velocity of the wind: You'll have elevator control much earlier, the takeoff distance will shrink in direct proportion to the wind's force, and when you reach the airport boundary your altitude will be considerably greater than usual; even though the groundspeed is reduced, the normal climb-out airspeed produces the same altitude gain per unit of time.

THE NORMAL LANDING

I have a friend who's one of a very small and exclusive group in aviation—he runs a successful small airport and he's not a flyer. (He says that keeping an aviator's emotions out of the flying business is the only way to make it pay!) Despite his lack of flight experience, he knows what's happening, as was evidenced one day by his remark on unicom to a solo student in a tricycle-geared airplane who had just made his third consecutive three-point landing. "Charley," my friend said calmly into the mike, "how many landing wheels are there on that airplane?" Charley didn't answer, but apparently he understood, because the next time around he had his act cleaned up and landed the Cessna firmly on the main wheels, then let the nose wheel—the *steering* wheel—sink to the runway as the elevators quit flying.

When you're talking about airplanes with the little wheel up front, my friend the non-flying FBO is absolutely right: The nose wheel is to be used for steering and for holding the front of the airplane off the ground while taxiing. That philosophy

Figure 16

The main gear should touch the runway first

leads to the definition of a normal landing: a gliding approach that terminates in a full-flap, power-off stall a couple of inches above the runway, and that results in the airplane's contacting the landing surface *on the main wheels first.*

A good normal landing is the aerodynamic opposite of a good normal takeoff, in which thrust moves the wing ever faster through the air and develops lift because you maintain an adequate angle of attack. On landing, you should once again put the nose on the horizon (or whatever other pitch reference you elect to use), but, with no thrust, the airplane will decelerate until the airflow over the wings is insufficient to generate lift equal to the weight of the airplane and you're on the ground. The trick, of course, is to make the stall happen very, very close to the runway—that's artistry in flying.

All good landings are preceded by good patterns, especially while you're learning. There's no better judgment-building exercise than the repetition of "closed traffic"—flying one rectangular pattern after another, with touch-and-go landings in between. If you are careful to arrive on the downwind leg every time at precisely the same altitude and airspeed, if you start your glide toward the runway from the same point with the same power and flap settings, it won't take many circuits for the sights and sounds and feelings to standardize. You'll learn what a good approach looks like, the effect of the wind will become very obvious, and you'll develop a solid base from which to make corrections.

A "good pattern" doesn't necessarily mean a rectangular pattern. The familiar crosswind, downwind, base, and final approach segments are great for training and for keeping everybody apart when the airport's busy, but in the real world you'll often be asked (or find it more convenient) to enter on base leg, to fly a dogleg to final or a straight-in approach—no "pattern" at all. Given the validity and practical application of these variations, a "good pattern" means a controlled approach to the landing surface, planned to meet the needs of a specific situation. Once the rectangle is mastered, work on improving your judgment so that you can enter from any point and still arrive on final approach in the same condition of flight (airspeed, altitude, configuration) that your training patterns produced.

Somehow, you must arrive "on final," lined up with the runway, and that's where the landing really begins. The old faithful cut-and-try system is the best way to find out what a good landing looks and feels like, so pick an airport with little or no traffic, a day with little or no wind, and start trying.

Establish a power-off glide at the same airspeed you used in the traffic pattern, then apply the first increment of flaps—one notch or ten degrees, whatever convenient measure applies to your airplane. Almost without exception, the nose will try to rise, but *don't let that happen*. Push just as hard or as much as necessary to keep the nose absolutely still pitch-wise, and trim

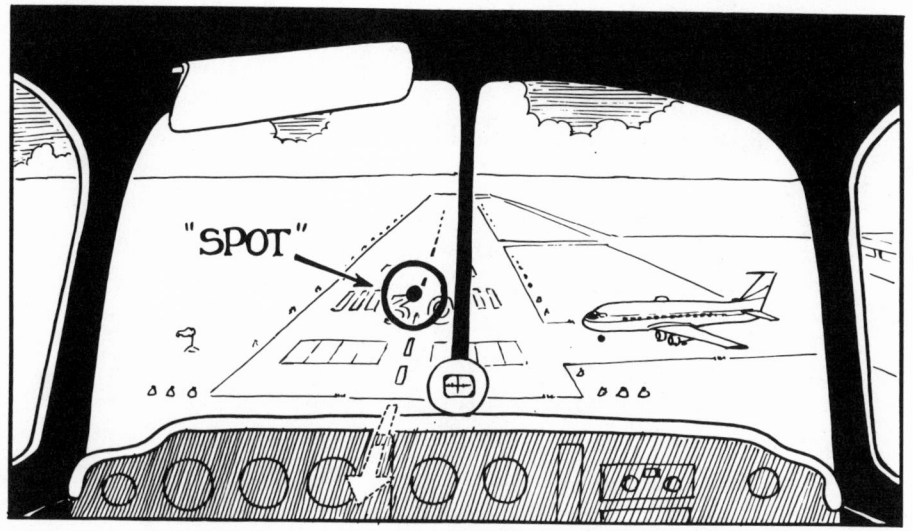

Figure 17

A normal approach viewed from the left seat of a typical side-by-side aircraft

for the new situation. The airplane will slow down a bit. When everything is stabilized, run out some more flap—you'll notice there is not as much trim change required to maintain the nose in the same attitude—and the airplane will slow down even more. By the time you have extended full flaps and allowed the airspeed to settle down whenever it will in that configuration, you should be able to tell whether you'll reach the runway. Once you've stabilized, there will be a spot on the windshield through which you are looking at the numbers on the end of the runway; if the spot stands still on the numbers, you're on the glide slope you want. Remember that you're using the spot as you would the front sights on a rifle, in effect aiming the whole airplane at whatever point you see through the spot. If it appears to stand still short of the numbers, that's where you'll land; a spot beyond the numbers indicates an overshoot.

Sooner or later, you'll come up with a combination that will produce a glide straight to the numbers. Obviously, if you do nothing the airplane will glide right into the ground nose-wheel first, so you must smoothly increase the pitch attitude (this is the *flare* or *roundout*) until the nose is on the horizon, and hold it there until the wings stall. No one can tell you when to start the rotation; the timing depends on the wind, aircraft weight, how rapidly you change pitch—enough variables so that the only way to find out is to deliberately rotate too soon on one approach, too late on the next. If you rotate too early and hold the landing attitude, the airplane will level off and stall more feet above the runway than you'd like. Should you wait too long to raise the nose, the little steering wheel up front strikes the ground first and throws the airplane back into the air. This maneuver is known among aviators as "le bounce bad." Having experienced when *not* to do it, start shooting for the in-between.

From an instructional standpoint, the area between the time you decide to begin the roundout and contact with the runway is a very gray area. Every increase in pitch attitude, no matter how small, slows the airplane a bit, which of course decreases the amount of lift generated by the wings at that particular angle of attack. The weight is unchanged, so when the lift drops off you've no choice but to increase the angle of attack again in order to keep the airplane from descending too rapidly. Now drag really starts to build up, and the resultant deceleration requires more and more pitch increase, until the wheel is all the way back in your lap—at which point the wings give up altogether, and if you've descended to just inches above the ground while all this is going on, the landing will be a "grease job." In effect, you've traded airspeed for lift, and there's a good bit of artistry involved; you must know when to start the flare, and once that is begun, you must figure out at what rate the pitch attitude must be increased. When the nose finally gets up to the horizon (or your personal reference point), and the airplane is reasonably close to the ground, the trick is to hold, hold, hold—maintain elevator pressure to keep the nose there. Even after touchdown, which in this attitude is almost guaranteed to happen on the main wheels, keep the wheel or stick all the way back—there's a lot of drag available from the flaps, the fully deflected elevators, and the belly of the airplane, and aerodynamic drag is the cheapest brake of all. When the elevators have bled the last ounce of lift from the rollout airspeed, the nose will settle to the runway all by itself.

All of us have an aversion to falling, and the undeniable fear of an imminent crash reinforces a pair of illusions that must be overcome if you're ever to get good at landing an airplane. The first is the apparent increase in the rate at which you're rushing toward the runway; the second makes the nose appear to drop as you get within a couple hundred feet of the ground. In both cases, Eyes tell Brain to think of something quick or we're gonna crash, so Brain tells Arms to pull back on the wheel, and a correction is made for a condition that never really existed. If you stabilized the approach earlier on final—full flaps, pitch attitude set and trimmed—there can be no change in rate of closure with the ground, and unless you deliberately move the nose up or down, the pitch attitude will remain the same, so the apparent drop is the result of more ground filling your field of vision as you get close to the runway.

These illusions are very compelling and must be overridden. One of the most effective ways to keep from being fooled is to look far down the airfield and out to both sides as soon as you feel the urge to haul back on the wheel; simply increasing the number and varying the sources of visual clues often solves the problem.

That's a normal landing. Even when a strong wind or a stretched-out pattern forces you to augment gravity with a little thrust from the engine, the actual landing portion of your return to earth should be a smooth transition from a power-off glide to a power-off stall just above the runway. Diligent practice at powerless approaches will make you vividly aware of the effect of wind on the flight of your airplane and improve your powers of aeronautical judgment, and should the engine quit in the middle of a flight one day, you'll be well prepared—there may not be a bona fide runway nearby, but all you have to do is turn into the wind and complete another of your now-familiar power-off approaches to the best-looking field you can find.

LANDING INTO A STRONG WIND

The techniques used to fly the airplane—the actual control pressures that make the airplane bank and turn, pitch up or down—don't change one iota when the wind velocity picks up. But the effect of that increased wind speed is sometimes remarkable, especially when you're dealing with lightweight airplanes. The slower the vehicle, the more it will be displaced by the wind. If you intend to stick with the power-off approach, you'll find some adjustments necessary to make everything come out the way you want it to.

For example, on the downwind leg of a rectangular pattern on a windy day, the groundspeed will be noticeably greater, providing you with less time to get ready—it will be time to start the turn to base leg before you can say GUMPS. If you power down and turn base over the same landmark that has consistently produced a beautiful final approach, don't count on it this time. The wind, now blowing from the side, will move your airplane away from the runway, and when you turn final you'll be located a lot farther from the threshold than a power-off glide can accommodate. There are two remedies: You might start the downwind-to-base turn considerably sooner, correct into the wind on base, and accept a slightly

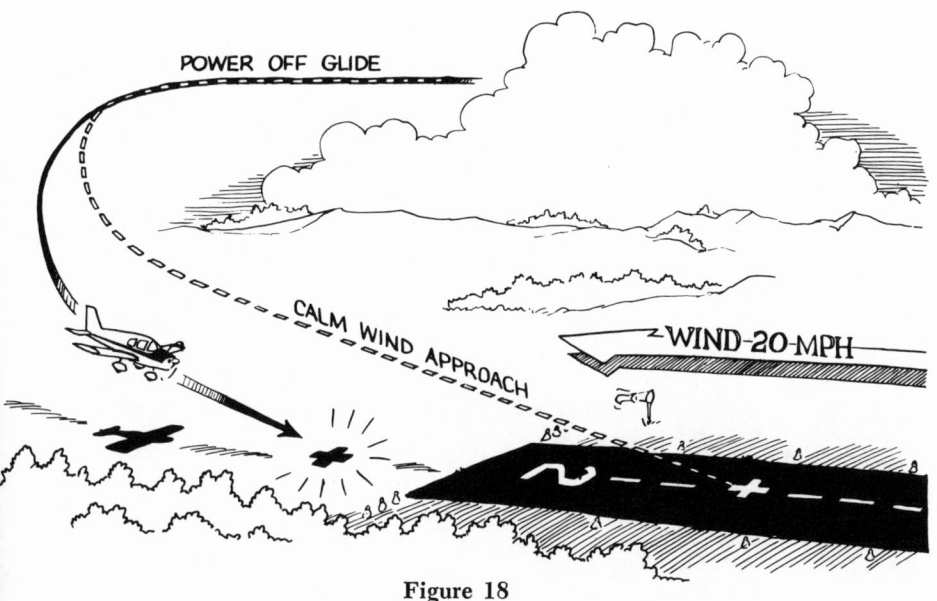

POWER OFF GLIDE

CALM WIND APPROACH

WIND-20-MPH

Figure 18

One of the most important skills in the art of aviation is
learning, anticipating, and correcting for the effect of the
wind

steeper glide path on final; or you can fly the no-wind pattern
across the ground and use engine power to make up for what
you'll lose to the wind.

The latter case is a beautiful demonstration of the difference
between *air miles* and *ground miles*. Imagine flying on a perfectly calm day, thousands of feet above the ground in a
power-off, full-flap configuration at 80 miles per hour; your
task is to discover how far the airplane will glide under those
conditions in one hour. Just for the sake of simplicity, assume
that you finally touched down 80 miles ahead—that would be
80 miles across the ground, and roughly 80 miles through the
air. Now climb back to the original altitude and do it again,
but this time with a 20-mph wind blowing in your face all the
way down. The time for the descent would be the same, but
you'd wind up 20 miles short of the original touchdown point.
You'd have traveled 80 miles through the air (airspeed × time

= distance flown in air miles), but the air happened to be moving in the *opposite* direction at 20 miles per hour. The amount of power needed to reach the no-wind touchdown spot would be exactly that required to carry the airplane 20 miles in still air—you'd have flown, in effect, 100 air miles (80 miles across the ground plus 20 miles' worth of wind effect).

The same thing will happen on a much smaller scale when you fly a normal pattern across the ground on a windy day. You will inevitably need power—the thrust requirement grows with wind speed—to get the airplane to the intended landing spot. And that's not all bad—at least you'll be flying approaches that always look the same from where you sit; they'll just take a little longer.

THE CROSSWIND TAKEOFF

Imagine your airplane parked on a huge frozen lake where the surface is as smooth as a mirror. With very little friction between ice and tires, forward movement begins almost as soon as the prop starts to turn, the airplane accelerates rapidly to liftoff speed, and your path across the ice is as straight as a string—*if* there's no wind.

Now introduce some wind blowing from one side or the other. With no corrective action on your part, the airplane turns into the wind just like the big weathervane it is—and so long as no one cares in which direction you take off (that's one of the nice things about flying off big lakes), why not *let* it turn, and take advantage of the direct headwind to improve performance?

More often than not, takeoff will have to be accomplished with the wind trying to push the airplane sideways. And push it will—on the near-frictionless lake surface, you'll find yourself moving downwind in direct proportion to the force of the wind, the tires sliding harmlessly across the ice. Even if the available takeoff area were limited to a long strip of ice and the wind were blowing across it, you could point the nose of the airplane into the wind enough to counteract the side forces during the takeoff roll. With almost no friction, nothing to keep the tires from skating smoothly over the ice, you'd have no problems with the crosswind force.

But when taking off from a runway paved with tire-grabbing asphalt or concrete, the sideways slide is out of the question. Even on an icy runway or a wet grass strip, where the coefficient of friction is low and the tires *will* slide readily, the usable takeoff area is seldom wide enough to allow lateral movement. So the objective for a successful crosswind takeoff is a simple one: to operate the controls so that the airplane rolls and then flies straight ahead with respect to the ground, the longitudinal axis parallel to the direction of travel at all times.

The technique to make this happen is equally simple, and consists mostly of the pilot's determination to use whatever's at his disposal to do two separate things: control any sideways movement with the ailerons and keep the nose on a point— straight ahead, of course—with the rudder pedals. A proper crosswind departure from the ground can take place just that mechanically, but smooth, confident application of the technique is enhanced by a complete understanding of what's going on.

To begin with, the weight of the airplane at rest creates so much friction between tires and runway that a sidewind strong enough to move the airplane would surely topple it. The downwind tire becomes a pivot point, and the airplane would roll around that point—hold it, there's the secret to crosswind technique, the fact that the airplane will try to *roll* when a side force is present. From the very start of the takeoff, as soon as there's any wind at all flowing over the ailerons, you have control forces to counteract that rolling tendency. How much is enough? Whatever it takes to keep the wings level. On a *really* crosswindy day, you might as well start out with full aileron, then take out what you don't need as the speed builds up and the controls become more effective. You are "flying" the airplane from the time the brakes are released, using the aerodynamic properties of the ailerons to keep the wings level. The increased lift on the downwind wing supplies enough rolling moment to overcome the force of the wind blowing from the side.

All goes well until the airspeed builds up and the weight of the airplane begins to transfer to the wings—now the friction between the tires and the runway decreases rapidly, and the force of the wind may well be enough to move the airplane sideways, usually a skip at a time. (If, on one of those skips, a tire gets a good bite of concrete, the momentum of the sideward movement plus the force of the wind pushing against the side of the airplane may cause an upset.) To maintain your track down the middle of the runway as the airplane gets lighter on its wheels, all you have to do is keep flying—in the finest sense of the word—and when the wind is strong enough to push the plane sideways before it's ready to leave the ground, use whatever aileron pressure is needed to stop the drift even if that much aileron displacement raises the downwind wing. There's absolutely nothing improper or unsafe about roaring down the runway on one wheel, and the added into-the-wind lift you are creating with this bank turns out to be just enough force to offset the wind's push.

That takes care of one dimension—sideways movement—but you must also deal with directional control during a crosswind takeoff. Discounting for the moment the nose-swinging effects of the propeller, consider only the tendency of the airplane to weathervane: It will always try to turn into the wind. The solution to the problem is quite simple: When the nose moves to one side or the other from the visual reference you're trying to hold, apply whatever rudder pressure is needed to stop the

CROSSWIND TAKEOFF

Figure 19

Control positions at the start of a crosswind takeoff

movement and bring the nose back where you want it to be.

When a crosswind is so vigorous that it requires aileron pressure to stop the drift, you are, in an aerodynamic context, banking the airplane, which will produce some turning in itself. You'll find that a strong crosswind will require *crossed controls*, with, perhaps, left aileron to stop drift, right rudder to hold the nose straight. A pilot no-no? Not on your life; you are doing precisely what you should be doing—keeping the airplane on a string-straight track down the middle of the runway, drifting neither left nor right—truly *flying* the airplane in a forward slip, even though you haven't left the ground.

Sooner or later, you must take care of the third dimension, pitch. In all but the very strongest crosswinds—at which point you should give some thought to using another runway, or maybe even delaying takeoff until the wind settles down or straightens out or both—the nose should be raised to the horizon as in a normal takeoff. Do it just as smoothly, just as firmly, but with the understanding that this increased pitch attitude will lighten the load on the tires sooner, and you'll have to accomplish the one-wheel, forward-slip-on-the-runway routine somewhat earlier in the takeoff roll.

When you finally part company with the earth (and, given the often gusty nature of crosswinds, you may have had a battle royal all the way down the runway), there are at least three things you can do: continue flying in a forward slip, level the wings and fly the runway heading, or set up a wind-correction angle (*crab*) sufficient to continue your track across the ground on the extended center line of the runway. The first is uncomfortable, inefficient, and increasingly difficult to accomplish as you climb, because the drift reference (the earth) is receding. The second is okay if where you want to go is downwind from the airport, but the "let 'er drift" technique (it's really no technique at all) has a lot going against it for safety reasons.

Which leads directly to a strong endorsement of number three, establishing a drift-canceling wind-correction angle as soon as the airplane leaves the ground. Not only does this allow comfortable, coordinated flight, it forces you to look outside to determine what crab angle you need (looking outside an airplane is *always* safe), it keeps you on a straight-out track during the initial climb (where controllers and other pilots expect you to be, and that's safe), and it provides maximum turn-around room should you come up with an engine problem and be forced to return to the airport. If all that isn't justification enough, look at it this way: Flying a track straight out from the runway is good aviation discipline; the airplane is doing what *you* want it to do. If you're going to be the pilot-in-command, be the pilot *in command*.

Every flying machine has its crosswind limits. When you have applied full aileron into the wind and full opposite rudder to keep the nose straight and the airplane continues to slide or skip toward the downwind edge of the runway, that's more crosswind than it's designed to handle. Of course, it's a hell of a time to find it out, when you're faced with a choice of skidding off the runway or letting the airplane turn into the wind and slide partly sideways down the runway with the very good chance that it will upset. Most airplane operating manuals state a *maximum demonstrated crosswind component,* which implies little more than how much of a direct crosswind someone in the flight-test program was willing to fly in; however, it's an indication of where a good pilot began to have doubts about controllability, and it's probably a condition that should be reserved for sometime in your future, when you've a lot more flying experience. Remember that when you have the controls fully displaced to overcome a crosswind, there's nothing left for any other into-the-wind corrections; it's all downwind from there, with the wind in charge.

Normal liftoff technique (raising the nose just as soon as the elevators are effective and allowing the airplane to fly off the ground by itself) will sometimes require full aileron and rudder displacement, which is certainly not a sin (that's what controls are for), but moving the wheel/stick and pedals all the' way to the stops has a stigma that some pilots just can't overcome. In this situation, it's possible to hold the nose deliberately on the runway with forward pressure to take advantage of the more positive nose-wheel steering until sufficient airspeed has been generated to *guarantee* that when the nose comes up, the airplane will leave the ground—now. And there you are, aloft in a machine with controls you are reluctant to operate to the limits, faced with landing in a crosswind that will probably require a full throw of those same controls. Maybe any crosswind that feels as though it's going to use up everything you're willing to give is too much.

THE CROSSWIND LANDING

Hold a mirror up to the previous discussion of crosswind take-offs and there's the technique for approaching and landing when the wind is blowing across the runway. The same rules apply, the same aerodynamic phenomena occur: You must contact the runway with the longitudinal axis of the aircraft parallel to the direction of movement, and there must be no drift to left or right.

If the wind is blowing strong enough to drift your airplane one mile sideways for each mile flown forward, you could set up a final approach one mile out, one mile over, and that would get you to the runway, but you'd still be moving sideways when the tires met the ground—that's no good.

Using the one-mile-out, one-mile-over example again, you might be able to rudder the nose to the proper wind-correction angle the instant before touchdown, but then the airplane would be moving across the ground at an angle—that's no good either.

Or you could fly down the final approach course with the proper crab angle and maintain it until a split second before ground contact, when you plan to bring the longitudinal axis in line with the runway—a good plan, but it takes great judgment and timing, the kind gained only by long experience, and if you blow it . . . well, the least damage you'd do would be to the tire treads, a great deal of which would remain on the runway from the resultant skid down the runway.

Now try it the *right* way. Stay on the extended center line of the runway with whatever crab angle is needed (discipline again) until you're ready to get serious about landing. With

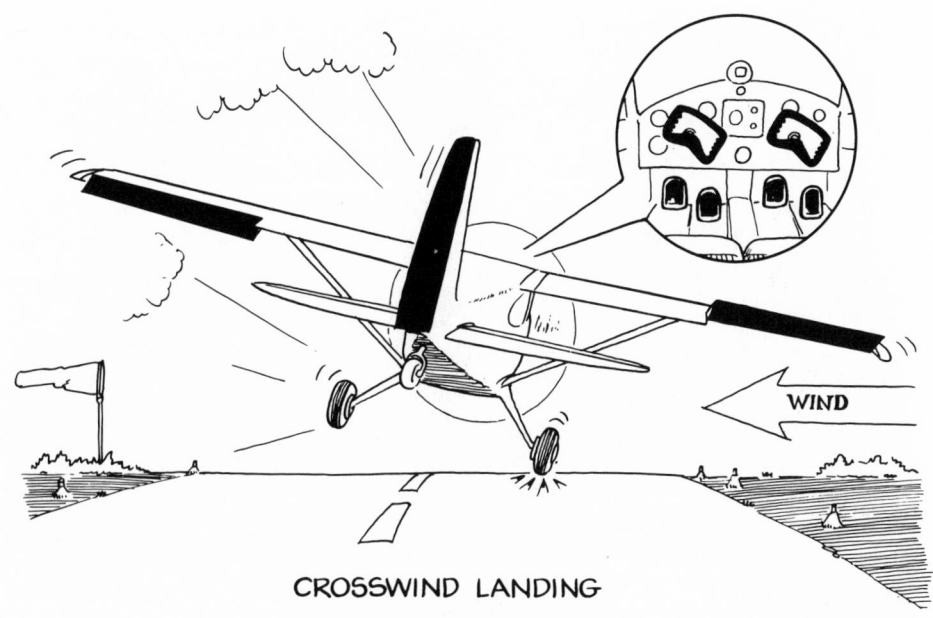

CROSSWIND LANDING

Figure 20

Control positions during a crosswind landing

the ailerons, bank into the wind so that the airplane is trying to move in that direction exactly as much as the wind is pushing ("enough" is when drift across the ground stops), and keep the nose-to-tail axis lined up with the runway, using whatever rudder it takes. You'll have crossed controls again, and you'll proceed down the center line in a forward slip, but there will be no sideways movement, and the wheels will roll straight ahead at touchdown.

At the same height above the runway as in a normal landing, slowly raise the nose to the horizon, increasing control displacement as necessary (as airspeed decreases, so will the effectiveness of any aileron, rudder, and elevator deflections) to prevent drift and to keep the nose on a point. Since you'll be slipping all the way through the landing process, expect to touch down on the upwind wheel first; then, as the speed continues to decay, allow the other wing to come down until its

---

wheel touches, and finally, when the elevators give up, the nose wheel should settle gently to the runway. From this point until you've got the airplane tied down, continue to keep the wings level with ailerons and the nose where you want it with rudder.

How about a little extra speed on a crosswind landing so you won't have to throw the wheel all the way over or push the rudder pedals to the stops? A little airspeed pad is okay, but remember that you've got to slow down sooner or later and do battle with that same crosswind; if you've got real doubts about whether you can handle it, you shouldn't be there. Go to an airport with a runway more nearly aligned with the wind.

One of the most significant problems associated with crosswind operations—both takeoffs and landings—is the recognition of drift. It never occurs when you're behind the wheel of an automobile (if the wind is strong enough to move your car sideways, it's time to head for the storm cellar!), and is very difficult if not impossible to observe in a boat, or on ice skates, or in almost any other form of transportation to which a beginning pilot is accustomed. But you can be trained to recognize drift, and the easiest and most effective way is to fly low over the runway with no wind correction and watch the airplane drift. When you can recognize what's happening, fly more low passes at progressively slower airspeeds and stop the drift using the techniques described above. When you are able to fly straight down the center line in the landing configuration at an airspeed not much above stall, can a good crosswind landing or takeoff be very far from your grasp? Not likely.

MAXIMUM-PERFORMANCE TAKEOFFS AND LANDINGS

Chances are better than good that the average light-plane pilot will never have the occasion, much less the need, to operate from a short field, one that severely taxes his or his airplane's abilities. Even with this low probability, it's good to know what maximum performance is all about, especially with regard to landings; you never know when you might have to put down in a farmer's field or a clearing in a forest that isn't long enough to accept the rollout distance you've been using back home on a long, paved runway. In general, the pilot who has some experience in maximum performance maneuvers in a training situation (with no obstacles other than imaginary ones, and plenty of runway ahead to accommodate the inevitable mistakes of a learner) will be able to do a much better job of flying into or out of a short field even if it requires only half the ultimate performance of the airplane.

"Maximum performance," just like "normal," needs to be defined: It's the technique of making the airplane lift off the ground in the shortest possible distance, then climbing at the steepest possible angle or rate until obstacles are cleared; on the arrival end of the flight, it's getting the airplane over whatever obstacles lie in the landing path and then bringing it to a full stop in the shortest possible distance. True maximum performance puts airplane and pilot at the very edge of their capabilities; it's advanced stuff, and shouldn't be tried without the guidance and helping hands (and feet) of a well-qualified flight instructor.

A maximum-performance takeoff is a variation on normal technique. Instead of allowing the airplane to accelerate through the "safe" speeds (best angle, best rate, normal climb) right after liftoff, you'll get right down to business, making the wing do its best work from the start. Put on your aeronautical engineer's hat for a while, because you must find out what power, pitch attitude, and flap setting will produce the greatest vertical speed. At altitude (enough to recover from an unintentional stall), set up full power (gear down, of course) and whatever flap setting the manufacturer recommends for a short-field takeoff, then raise the nose a little at a time until you discover the pitch attitude that produces the highest reading on the vertical-speed indicator. (Don't have one of those? No problem. Just hold each successive attitude long enough to time the increase in altitude and convert your readings to feet per minute. It helps to have someone along to write down the numbers.) When that point is reached, note the airspeed and the pitch attitude on both the attitude indicator and the real horizon—there are times when one or the other is more meaningful.

Back on the ground, with plenty of information and confidence that the number and the attitude you discovered will work, put the flaps where they should be, hold the brakes, and run the engine up to full power. The period of time to accelerate from brake release to the best-angle-of-climb airspeed—that's what you derived from your experiments—will probably surprise you; it doesn't take long. As the needle on the airspeed indicator approaches the magic number, smoothly but firmly raise the nose to the predetermined attitude and hold it there, and the airplane will come off the runway and climb just as rapidly as it's able to. Short takeoff run, high rate of climb—that's maximum performance.

There is a technique within this technique, and it deals with the fact that a too-soon displacement of the elevators to raise the nose creates enough drag to hinder acceleration. This effect is less noticeable on higher-powered airplanes, but is bound to steal some performance no matter how many horses are under the hood. After a bit of practice, you'll know when to start the nose up and how fast to move it. If the pitch attitude is increased smoothly and continuously and the airplane comes off the ground just as the nose reaches the desired attitude, you've done everything right. Practice may not make it perfect, but it sure helps.

Once you're airborne, the situation at hand controls your next move. If there's an obstacle very close by that needs to be climbed over, hold the airplane in the maximum-performance attitude until you're clear. When the problem is just a very short airfield—no obstacles—it makes a lot of sense to lower the nose a bit to obtain best-rate-of-climb airspeed as soon as you're firmly supported by the atmosphere.

If a maximum-performance takeoff is questionable for the beginner, a maximum-performance landing is twice as questionable. Everything has to work just right, and the really short fields for which you'd need this kind of performance are usually unpaved, with plenty of obstacles all around, and are located in out-of-the-way places—likely as not, in the mountains, where performance altitude is bound to work against the efforts of even the best pilots.

There are two types of approach/landing techniques involved in getting an airplane down and stopped in a very short distance, just as there are two types of normal approaches and landings: those in which power is used to control the glide path, and those in which the landing comes at the end of a powerless glide. The power-off method is good for developing judgment and serves as a fine drill to sharpen your forced-landing procedures, but when you are given the luxury of choosing, the powered approach should be used to extract every last bit of landing performance from the airplane.

For practice, set up a longer-than-normal final approach and configure the airplane as it will be at touchdown: wheels down and flaps fully extended. Now find a pitch attitude that will produce an airspeed just above stall; here's where you should add a few knots or miles per hour for the wife and kids, or for yourself if that's your primary concern, because any gustiness in the final-approach wind will make subtle changes in angle of attack, and you should have enough of an airspeed cushion to accommodate them. Once the attitude-airspeed combination is determined, maintain the nose right there and use the throttle as your sole source of glide-path control—add a little to decrease the rate of descent, take away some power if it appears that you're going to overshoot. Since you're dealing with judgment and a host of visual illusions, the only way to get good at adjusting the glide path is to practice until you know what a proper approach looks and feels like. At the heart of the procedure is a constant-pitch attitude. When you master that, airspeed won't vary throughout the approach, nor will your reference points on the horizon.

The attitude that produces the desired airspeed will also turn out to be the desired landing attitude for most light aircraft. Therefore, if the glide path is controlled so that you arrive at the touchdown point in the landing attitude, there will be no roundout, no flare, and no float; the airplane should contact the ground main wheels first, positively and firmly. "Grease jobs" and "maximum-performance landings" are not necessarily synonymous.

Once on the ground, there are two kinds of braking available; *aerodynamic*, produced by the drag of full flaps, fully deflected elevators, and the increased frontal area of the whole airplane in its nose-high attitude, and *wheel-braking*, which is most efficient when applied early in the landing roll and when the coefficient of friction between tires and runway is increased by retracting the flaps and transferring more of the aircraft weight from wings to wheels. The situation in which you find yourself should dictate your braking procedure. Unless the runway is super-short, hold the nose up and leave the flaps where they are—aerodynamic forces are better brakes than you think, especially when landing into a strong wind. Of course, when you need all the stopping power you can muster to keep from going off the end, get the flaps up and press on the brake pedals right away. Pumping won't help; apply steady, increasing pressure until the airplane comes to a stop.

Even more than a maximum-performance takeoff, a properly executed approach and landing to a short field requires careful management and control of the four forces of flight. Occasional practice of these techniques will help to develop a reservoir of experience, an in-depth knowledge and understanding of the capabilities of both aircraft and aviator; and should the unfortunate situation ever develop in which you need that knowledge, the far corners of your airplane's performance envelope won't be quite so mysterious. The noticeable improvement in your normal landings and takeoffs is but another bonus to be earned from your efforts at maximum performance.

Flight Instruments

6

The behavior of an aircraft in flight is very predictable: Put it in some predetermined attitude, apply a certain amount of thrust (or take some away), and it will climb, descend, fly level, go faster, slow down, or do whatever that combination of attitude and power dictates. The power-setting part is easy; establishing a predetermined aircraft attitude and insuring that it's the same every time are a bit more difficult. Equally important are the *results*. Are you flying at the speed you aimed for? Are you maintaining the altitude you want in level flight? How much farther should you descend before leveling off? The answers to those questions and a thousand others encountered on every flight make up the evaluation of your efforts. The answers are right there in front of you, displayed on the flight instruments.

Now there's a lot to be said for seat-of-the-pants flying—after all, that technique stood alone for many of aviation's early years—but until a pilot has pressed his pants into the seat of the same airplane for a lot of hours, the performance feedback from his posterior is rather unreliable and grossly inefficient. You can do a lot better with even the most unsophisticated, bare-minimum instrumentation found on nearly all light planes today. Not only are basic flight instruments a big help, they are required by law on any airplane you'll use for pilot training. And since, as part of the certification flight check, you'll be required to demonstrate the ability to control your airplane on instruments—i.e., with no outside references—the more you know about the tools, the better job you'll do.

This chapter won't tell you how to fly on instruments, but rather how the instruments operate. You need to know what goes on behind the glass faces on the panel because there are unavoidable errors in almost every one of them—errors that can become insidious traps when they're not understood. The flight instruments can be grouped according to the natural forces on which they depend for their operation: magnetism, air pressure, gyroscopic forces. Even the technological wonders of avionics are for the most part just better ways of presenting to the pilot the basic information that the instruments sense.

THE MAGNETIC COMPASS

Old Reliable, Ever Faithful, Always Ready: The magnetic compass is the tortoise of the instruments, having plodded faithfully through navigational history in much the same form as you'll see it on the most modern aircraft. It's the same instrument that guided Columbus to America, countless Boy Scouts across trackless wastes, and Lindbergh to Paris. High-performance aircraft with much more sophisticated direction-indicating systems carry a lowly magnetic compass tucked away in an obscure corner of the instrument panel; their crews invariably refer to it as the "standby" compass. Its dependability is legend, its information a navigational last resort, for the compass senses the natural force that is perhaps the most fundamental of them all: the Earth's magnetic field.

Differing from the one you get in a Cracker Jack box only in its adaptations for aerial use, a magnetic compass has two iron bars that always remain parallel to the invisible lines of force that surround the Earth. The bars are fastened to a float-mounted card, which pivots on a jeweled bearing, the whole assembly nearly weightless and friction-free. Because you can see only a portion of the numbered card, it appears to turn when you change the airplane's direction, but not so—the iron bars maintain their position of alignment with the magnetic field, and the airplane turns around the compass card. Since you're interested in which direction you're headed, the card is arranged so that "N" appears when the airplane is headed north, "S" when you're headed south, and so on around the directional circle, with every thirty degrees numbered, every five degrees indicated by a vertical line.

Five degrees is about as close as you'll ever come to navigating accurately with a "whiskey" compass (nicknamed for the fluid in which the card floats, usually a low-freezing-point alcohol base), because every bump in the air makes the compass jump around; the bigger the bumps, the wilder the jumps. And every time the compass jumps, it swings a bit, making it somewhat less than easy to read in anything but smooth air. What to do? Average the readings, use the number you see at the midpoint of each swing, and you won't be far off. That's *oscillation error*, logical, easy to understand, and predictable.

A different kettle of magnetic fish is *deviation error*. It's one of the insidious ones, almost always present, but in relatively small quantities and subject to change without your knowledge. Anything within the aircraft that generates a magnetic field will influence the iron bars in the compass, and although they may be trying to show you the proper direction, they're deflected from their purpose by stray magnetic forces.

Aircraft are built mostly of aluminum and other nonferrous metals for the sake of weight, but there's always some steel in the structure, some of it close enough to the magnetic compass to affect its accuracy. Of greater importance is the magnetic field created around each wire through which an electrical current passes, and there are always plenty of those up front. The compass cares not about the source of magnetic fields; it can only sense the effect. Whenever forces other than the Earth's are introduced, an error will be present, and it's called *deviation*.

An individual-airplane error, deviation will often have a number of values for each installation, depending on the combination of radios, lights, and other electrical equipment in use. To arrive at a correction factor that's somewhere near accurate, the airplane is placed on a compass rose painted on the airport ramp, with lines of predetermined accuracy radiating in the cardinal directions. With the engine running and the radios on (that's the configuration for most flights), the airplane compass is calibrated with reference to the painted lines. Errors are seldom very large; when the difference between actual heading and the reading on the instrument grows much beyond five degrees, it's time to have someone take a look at the compass or the wiring, or both.

This procedure is called *swinging the compass,* and when it's completed, the differences are noted on a compass-correction card, which becomes part of the airplane's paperwork. Usually mounted close by the compass itself, the correction card lets you know what number should show up under the lubber line when you're actually pointed in a desired direction. For example, turn the airplane until 276 appears, at which time the actual magnetic heading will be 270 degrees (see Figure 21). It's unlikely that six degrees will concern you that much in everyday flying, but deviation error can be the source of a great deal of navigational confusion when compounded by a series of bad readings.

So much for the small errors. The ones that can cause real

FOR	N	30	60	E	120	150
STEER	006	33	69	99	131	162
FOR	S	210	240	W	300	330
STEER	190	220	250	276	306	338

Figure 21

Compass-correction card

trouble are *turning* and *acceleration* errors, and they spring from a common phenomenon: *magnetic dip.*

Picture the earth as a huge magnet, with lines of magnetic force (often called *lines of flux*) entering at one pole and exiting at the other. The lines are parallel to the Earth's surface at the equator and perpendicular to the surface at the poles. In between, there's a constantly increasing vertical component, and the north-seeking ends of the compass bars will be pulled downward, in addition to lining up horizontally with the magnetic field. The farther you travel from the equator, the more pronounced this vertical effect becomes. If nothing were done about it, the iron bars would respond to the downward force, and you'd be looking at the underside of the compass card. But inside the compass housing are compensating magnets, which can be adjusted to apply enough upward pull to exactly counterbalance the vertical component of the earth's magnetic field; unless you fly very far north or south, the card in your compass will look you straight in the eye.

So long as the airplane (and therefore the compass card) is level, the compensated vertical force has no effect, but when you're headed north or south and tilt the card, it will rotate—that's *turning* error. Even though the aircraft heading hasn't changed one degree, the compass indicates a turn. The steeper the bank, the faster the card will turn, but in general the amount of rotation is dependent on the aircraft's latitude (directly related to the amount of vertical force) at the time; the compass will turn roughly the number of degrees that corresponds to your position north of the equator.

The vertical force will act whenever the card is tilted, which means that whenever the aircraft is banked the compass will read improperly. Level the wings and the card swings back to its former, proper reading. Turning error will be most noticeable when you are headed either north or south, because the compass bars are lined up with the nose-to-tail axis of the airplane and therefore have farther to "fall" when the card is tilted. For the same reason, turning error is nonexistent on headings of east and west.

Because you're looking at the back of the card (otherwise you'd see where the compass is pointing instead of which direction you're headed), the compass indicates a turn in the direction *opposite* the bank you've established. Roll to the right (as if to turn to the right) and the compass indicates that you've turned left. Should you indeed enter a turn, the compass will gradually catch up with the actual heading of the airplane, and should read properly as you turn through east or west. It's that initial surge of incorrect indication that causes the trouble.

The same situation—tilt the card and the compass turns—prevails on a heading of south, but because of the way the card is numbered, it will appear that you are turning in the *same* direction at a much faster rate. Once again, the error will show up whenever the airplane is banked, and if you continue with an actual turn, the card will indicate the proper numbers as you pass through east or west.

As if to defy an aviator's efforts to find out which way he's

---

headed, the magnetic compass goes haywire under another set of circumstances when the airplane is pointed east or west (up till now, those headings have been error-free). The problem stems from the same troublesome vertical component of magnetic force, but occurs only when the airplane is accelerated or decelerated. Picture the iron bars when the airplane is headed east: They're still lined up with magnetic north, but are now perpendicular to the nose-to-tail axis of the airplane. Being somewhat underslung on its jeweled pivot, the compass card will react to acceleration by tilting, the pulling-down force will cause the tilted card to rotate a bit, and the numbers will indicate a turn even though the airplane hasn't changed its heading at all.

The magnitude of this error is largely determined by how rapidly you speed up or slow down, and so long as acceleration or deceleration exists, the compass will read incorrectly. Headed in either direction (east or west), the compass will indicate a turn toward north when you accelerate, a turn toward south when you decelerate. Resume a steady speed and the compass drifts back to where it belongs. The secret is to know what's happening and restrict your compass readings to those times when airspeed is steady.

Since a magnetic compass is reliable only when it's level and not subject to the forces of acceleration, you may wonder how Columbus managed to find the New World—a ship races down the side of one swell and crawls up the next, rolling all the while. Check a marine compass and you'll find it mounted in gimbals so that the ship can roll on its beam ends and the compass stays upright—a simple solution for sailors, but more complicated, costly, and heavy than we aviators can justify. A pilot must be aware of the unavoidable errors in his compass and believe it only in wings-level, unaccelerated flight.

Every airplane must have a magnetic compass—that's the law—but the reliability of gyroscopic heading instruments has relegated magnetism to a standby, use-it-if-all-else-fails role; the fine points of outguessing its errors have become a lost art. And perhaps that's just as well—trying to follow the gyrations of a bouncing, swinging magnetic compass is further proof that the old days were not necessarily good.

THE AIR-PRESSURE
(OR LACK THEREOF) INSTRUMENTS

The generality of their forecasts notwithstanding, there are some people who can tell you when bad weather is on the way by the feeling in their bones. And have you noticed that folks are happier when high pressure dominates the weather scene, how depressed they become when a low-pressure system prevails? We've become so accustomed to living on the floor of an ocean of air that our bodies and minds respond readily to changes in atmospheric pressure. These are passive pressure experiences, brought about by the constant movement of weather systems across the Earth, and, of course, we have no control over the magnitude and frequency of the changes.

With pilots, it's a different story. In addition to the changes everyone else feels, each flight involves a vertical excursion through the atmosphere and an active encounter with air pressures that change frequently and significantly. Pressure changes are at once the inevitable result of an airplane's movement through the atmosphere and the source of information that lets the pilot measure those horizontal and vertical movements.

A pair of flight instruments that sense changes in air pressure provide reasonably accurate information about altitude and airspeed, and their names are about as straightforward as you can get: the *altimeter* (*altitude meter*) and the *airspeed indicator*. Note that they supply *reasonably* accurate information; just like the magnetic compass and its sometimes strange ways,

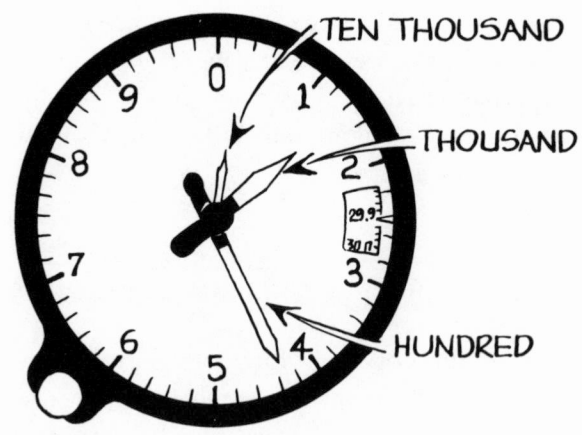

Figure 22

Altimeter face

the altimeter and airspeed indicator don't always tell the truth. Fortunately, air-pressure-instrument errors can't cause serious problems for the fair-weather pilot, but a little knowledge goes a long way toward more efficient operation . . . and a little aviation knowledge is seldom a dangerous thing.

HOW HIGH ABOVE WHAT?

The altimeter is a portable barometer adapted for use in airplanes, and it's a relatively uncomplicated instrument. A system of sealed bellows responds to even the smallest change in air pressure by expanding or contracting; the movement of the bellows is magnified by a gear train, and the result is displayed to the pilot on a scale marked in feet. Since atmospheric pressure is related to the weight of the air above, it stands to reason that pressure will decrease with altitude. The altimeter is mechanically arranged so that whenever the bellows senses a decrease in air pressure, the pointers on the instrument panel show an increase in altitude, and the other way around.

On a day when the air pressure at the surface of the Earth is uniform and standard, all altimeters at sea-level airports indicate just that—sea level, or zero feet. Each of these altimeters indicates 1000 feet when lifted far enough into the atmosphere for the bellows to expand and drive the hands to the 1000-foot mark. Bring them back down to sea level and the bellows senses standard pressure and causes the pointers to show zero feet once again.

If Torricelli (inventor of the barometer) had strolled up a mountain with one of his gadgets in hand, he'd have discovered that the mercury dropped just about 1 inch for each 1000 feet he climbed—and that's the pressure-altitude relationship built into the altimeter. The smooth, continuous decrease of air pressure during a climb is sensed by the bellows, which expands smoothly and continuously and causes the pointers to indicate a smooth, continuous increase in altitude. At level-off, the altimeter hands stand still, because there's no pressure change, but when you start down the bellows begins to contract and causes the pointers to indicate the descent. *For each 1-inch pressure change, the altimeter shows an altitude change of 1000 feet.*

All altimeters are supplied with standard-day sea-level bellows, so altitude discipline in aviation becomes a matter of deciding how many feet above sea level you'd like to fly, then climbing until the pointers indicate you're there. The height of all terrain and man-made obstructions to flight is published in feet above mean sea level (MSL), so if you want to make sure your flight over Manhattan won't tangle with the Empire State Building, climb until the altimeter reads at least 1250 feet. You'll have to go to 20,320 feet MSL to traverse the entire United States in complete safety; anything lower than that might run you into Mount McKinley. In any event, you'll understand the basic concept of the altimeter if you understand this: Altimeters measure air pressure, and display it in terms of feet above mean sea level on a standard day.

So far, no problems. But consider the situation in which lower-than-normal atmospheric pressure prevails over a sea-level airport (the kind of a day when Grandpa's bones really ache, and everybody you meet is grumpy as hell); the altimeter senses lower pressure, and shows a higher altitude. If the pressure happens to be, say, 1 full inch lower than the 29.92-inch standard, your altimeter reads 1000 feet, even though the airplane is still at sea level. When standard-day conditions return to your airport, the bellows are compressed, driving the pointers back down to the correct reading.

Atmospheric pressure is very seldom exactly 29.92 inches of mercury, and when it gets there it doesn't stay there long. But by turning the adjustment knob on the altimeter in the appropriate direction, you can move the pointers and cause them to read the proper altitude, just as in setting a clock to the correct time. On the nonstandard day just mentioned (pressure 1 full inch lower than standard), you'd turn the altimeter hands down 1000 feet, whereupon they'd read sea level and you could fly around the airport always knowing your altitude above the field. Every time the pressure changed a bit (in truth, it's constantly changing), you'd need to land and reset the altimeter to be sure the indication was correct.

Whenever you move the adjustment knob, there's a proportionate movement of the *altimeter setting scale*. It's marked in inches of mercury, and changes 1 full inch whenever you move the altimeter hands 1000 feet. Referring once again to a day when the pressure is 1 inch lower than standard, the altimeter setting scale would read 28.92 inches when you turned the hands back down to sea level. That's the same number a tower controller offers as the altimeter setting, a standard piece of aeronautical information that lets all pilots in a particular area set their altimeters to the same pressure reference. The altimeter setting is available from any Air Traffic Control facility; it's the pressure (in inches of mercury) that would exist at imaginary sea level at the airport. On the nonstandard, 1-inch-below-normal day, the pressure at the sea-level airport *is* 28.92, and when you move the scale to read 28.92, the altimeter hands indicate zero feet.

So there are two ways you can be sure the altimeter is reading correctly: Set the hands to indicate the height above sea level of a known point (such as the published elevation of the airport, or the top of Mount McKinley) or insert the current altimeter setting. Then, should you fly from the sea-level airport to Albuquerque and apply current altimeter settings all the way along the route, the altimeter will read 5352 feet when you land—no matter how much or which way the pressure has changed en route.

Measuring the vertical dimension of flight is very important: Pilots maintain a safe distance above obstacles by observing their published heights above sea level and comparing that to the airplane's height above sea level as indicated on the altimeter. You are able to make every traffic pattern the same by adding 800 feet to the field elevation and flying the resultant number on your altimeter. By observing the altitude rules on a cross-country trip, you can be reasonably certain that you'll have no less than 1000 feet of vertical separation from the guys going the other way—if everybody is using a current altimeter setting and observing the rules of the road.

Early-day altimeters had just one pointer, and no pressure-adjustment knob. That was good enough for those folks who never flew very high in the first place, could usually tell how high they were by the relative size of trees and houses and animals, and had no need for altitude accuracy. But the ability of today's lightest planes to fly at 10,000 feet plus, and the need for altitude standardization demand a more accurate indication —hence the three-pointer altimeter, with which most airplanes are equipped.

The longest of the pointers makes a complete circuit of the

altimeter dial whenever you climb or descend 1000 feet. The next-longest hand is merely a counter to help you keep track of how many times the long pointer has revolved; starting from sea level, both hands are on zero, but when you've climbed exactly 1000 feet the shorter pointer has wound up to the number "1" on the dial. It points to "2" at 2000 feet, and so on. It doesn't wait to move until the long hand is almost full circle; it moves a tenth of the way at the first hundred feet, and will be halfway to "1" at 500 feet, always moving in the proportion of 1:10.

When you reach 10,000 feet, the shortest hand gets into the act. It's been creeping upward all the time at the same 1:10 rate, counting off tens of thousands of feet. The pointers are linked mechanically, and the ratios never change.

In a low-performance airplane, altitude never changes so rapidly that the pilot should be confused about his height; you should always have a rough idea of your altitude even without looking at the altimeter. But to stay out of trouble, force yourself to read the altimeter from the inside out: 10,000-foot pointer first, then the 1000 foot hand, finally the 100-footer, which on most instruments can be read to the nearest 10 or 20 feet. You say you never fly anywhere near 10,000 feet, certainly never above that altitude, so why have any concern for the shortest pointer? That's okay if you never go near high mountains in your airplane, but on that first flying trip across the Rockies, when you must fly higher than you've ever flown before, don't be surprised if your altimeter fools you now and then. Play it safe—form a good habit at the very beginning.

Speaking of good habits, don't neglect the source of static air pressure that feeds data to the altimeter and other air-pressure instruments. It must be checked before each flight to make sure the holes aren't plugged or deformed. It's very embarrassing to leap into the sky and have to guess your altitude and airspeed throughout the flight.

There must be a better way to indicate such an important thing as altitude, and there is. Easier-to-read altimeters are available, most of them with digital counters instead of the two smaller pointers. In a typical presentation, a single long pointer revolves once for each 1000 feet, and a number rolls into view to let you know how many times the pointer has gone around the dial. Reading this type of altimeter is a matter of noting the lower number on the digital counter (since it moves proportionately with the pointer, there will be parts of two digits visible nearly all the time) and adding to that the number of feet indicated by the pointer. For a lot more dollars, you can get an altimeter with a pointer plus counters that show exact altitude, just like the odometer on your car; instead of miles, it counts feet above sea level. The ultimate is a fully digitalized presentation with no pointers, no counters, just a lineup of numbers that tells your altitude with no interpretation required. Such altimeters are already installed in more sophisticated airplanes, and eventually they will find their way into the general aviation fleet.

Altitude expressed as feet above mean sea level is just one of several definitive types of height measurement used in aviation; the other kinds of altitude will be discussed later on, in as much detail as they warrant. For now, concern yourself with the one kind of altitude that's most important in any flight situation—enough altitude.

AIRSPEED INDICATOR

How fast does a kite fly? On a windless day, a kite "flies" at whatever speed the boy at the end of the string can run at. On a breezy day, when he stands still and lets the wind do the work, the kite "flies" at whatever speed the wind is blowing. In either case, it's the movement of the kite relative to the air that makes the difference between flying and fluttering to the ground—in other words, speed through the surrounding air; in another word, *airspeed.*

Put your hand out the window of a parked airplane and you'll feel no pressure—the atmosphere is pushing equally all around. Stick your hand out the window at 100 miles per hour and you'll likely wind up with a broken arm. The area of your hand hasn't changed, but the force of the relative wind has, resulting in greatly increased pressure acting on the forward side.

The principle of moving-air pressure is put to use in the airspeed indicator. Instead of measuring the changes in static air pressure caused by moving vertically through the atmosphere, the airspeed indicator takes into account the changes in *dynamic pressure* of the relative wind. This pressure is sensed by the *pitot tube,* which is located in a spot that the airplane designer has found to be free of disturbances in flight. An opening in the pitot tube faces directly into the relative wind. The pitot tube transmits dynamic pressure to the airspeed indicator, where it quite literally blows up a bellows that is almost identical to the one in the altimeter. As the airplane goes faster (or as the wind blows with more force on the ground), the bellows expands and drives the airspeed pointer to higher numbers on the face of the instrument. It makes absolutely no difference to the indicator where the pressure comes from, whether moving airplane or moving air. When it senses a certain pressure, it points to a certain number.

The case of the airspeed indicator is connected to the same line that supplies static pressure to the altimeter, so the reading on the airspeed instrument represents the difference between static and dynamic pressures—static on the outside of the bellows, dynamic inside. When the airplane is at rest on a windless day, there is no pressure differential, and the indicator should read zero (some indicators are "zeroed" when the pointer rests on some other number). As soon as you start moving down the runway, pressure begins to build in the pitot system and the bellows expands and makes the pointer move. Suppose that when takeoff speed is reached, the dynamic pressure is 2 inches greater than static; the pressure differential would cause the indicator to read a certain airspeed. When an altitude of 1000 feet is reached, the static pressure has dropped roughly 1 inch, taking some of the pressure off the outside of the bellows, but dynamic pressure has decreased the same amount, and the differential is essentially unchanged.

This relationship holds true throughout the low-altitude, low-airspeed environment of most light planes, but when you fly more than a couple of thousand feet above sea level, a discrepancy begins to show up between what the indicator shows and the *true* airspeed. There are temperature factors involved in the precise computation, but, as a rule of thumb, you can increase the airspeed you see on the instrument by 2 percent for each 1000 feet of altitude above sea level. When the pointer indicates 100 knots at 7000 feet, your true airspeed is very close to 114 knots—nice to know when you're figuring time estimates and fuel reserves.

Nearly all indicators are ringed with colored bands to remind the pilot of several airspeeds important to safe and efficient operation. The color-coded airspeeds include a correction for errors within the instrument itself, but in small airplanes there's usually not enough difference to matter. Consider those color bands as a listing of performance airspeeds wrapped conveniently around the outside of the indicator and you can't go wrong. Here is the code:

WHITE BAND: The low-speed end represents the power-off stalling speed of the airplane in its landing configuration (wheels extended, wing flaps all the way down) at its maximum gross weight. The high-speed end of the white arc has nothing to do with stall speed, but is the highest airspeed at

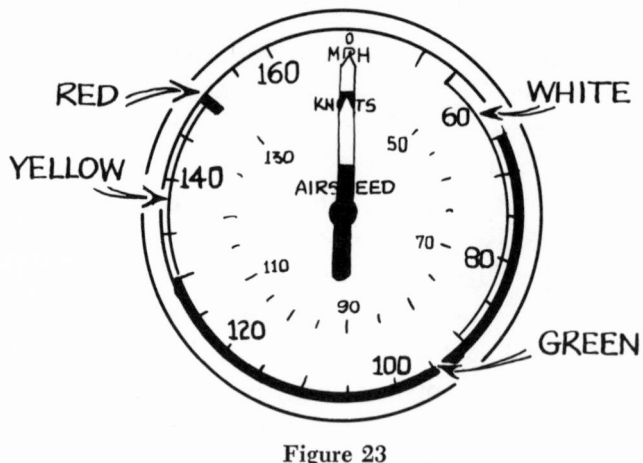

Figure 23

Airspeed indicator

which the airplane should be flown with the wing flaps fully extended—anything faster with full flaps and the manufacturer's guarantee of structural integrity is null and void. Nearly all airplanes have higher permissible airspeeds for partial flap extension. Know what those numbers are for your airplane, as partial flap settings provide fine speed brakes in certain operational situations.

GREEN BAND: Usually begins just a few knots higher than the bottom of the white arc, and the low-speed end indicates the speed at which you should expect the airplane to stall when it is "clean"—gear and flaps tucked neatly away—and no power is being applied. Like all the others, this speed has been derived from flight tests with a fully loaded airplane, a situation that you'll probably never encounter, since you'd have to have taken off at a weight greater than maximum allowable, and that's a no-no. The color-coded stall speeds therefore provide a "safe" error; the airplane will almost always fly at speeds considerably lower than those indicated on green and white. The green arc stops at a rather nebulous number; it's known officially as *maximum structural cruising speed*, which is as fast as you should allow the airplane to travel in rough air or when you intend to really horse it around with full displacement of the controls. Where the green band stops is not nearly so important as where the yellow band begins.

YELLOW BAND: The airspeed range covered by the yellow arc is known as the "caution" range, and you've no business flying at those speeds unless the smoothness of the air is surpassed only by smoothness of your control inputs. It's a matter of structural integrity, because each knot of speed contributes to the effective weight of the airplane whenever you enter a turn in level flight or pull out of a dive. (Remember centrifugal force?) Most light planes are incapable of airspeeds in the caution range when in level flight—you'd have to put yourself in a very uncomfortable situation to get into trouble. Higher-performance airplanes are frequently flown at yellow airspeeds in level flight, and almost always enter the caution range when descending—but should the air get rough, a smart pilot slows down into the green.

THE RED LINE: Easiest of all to understand. It's the highest speed at which your airplane was designed to be flown—period. Go faster than that and if the airplane comes apart, don't expect to recover from the guy who built it. The manufacturer's test pilots have proved the airplane will do it, so it isn't necessary for you to make sure it will go that fast. Anyway, you wouldn't like the sights and sounds unless you're a frustrated dive-bomber pilot.

Unfortunately, the airspeed-indicator builders have left unmarked the two most important speeds: best rate of climb and best angle of climb. You'll undoubtedly have to recite the antiquated "Litany of the Airspeeds" when you are examined for a pilot certificate, but why not mark those two speeds on the indicator, where they'll do you the most good? Go right ahead—there's no law that says you can't mark your own instruments. A narrow stripe of bright paint or a thin strip of contrasting pressure-sensitive tape gives you a beautiful target, one you don't have to call up from memory. Want to climb at best rate? Open the throttle and adjust the pitch attitude until the airspeed needle points to your best-rate mark. When you get into a crack and need all the altitude you can get *right now,* time-to-climb be damned, pour on the power and pitch up until the needle rests on the best-angle airspeed. You really shouldn't care whether your target is a number or a mark on the glass, so long as that power-attitude combination produces the performance you need.

The last of the air-pressure instruments is the least important. Little more than an altimeter with a carefully designed leak in its bellows, the *vertical-speed indicator* shows the rate at which altitude is changing. Should you climb suddenly for several thousand feet, then poke a hole in the bellows of your altimeter, the bellows would collapse until pressure inside was equal to pressure outside. Therein lies the principle of operation of the vertical-speed indicator. As pressure changes (in climbing or descending), the "leaky altimeter" tries to match inner and outer pressures, and displays rate of change to the pilot as feet per minute up or down. Since there are no flight maneuvers that require a specific rate of climb or descent, vertical-speed information falls into the "nice to know" category.

THE GYROSCOPIC INSTRUMENTS

Get anything spinning fast enough and it becomes a *gyroscope.* When that happens, certain laws of physics take over; namely, the gyro will remain stabilized in its plane of rotation, and an attempt to turn the gyro will cause it to tilt. These two principles, *rigidity in space* and *gyroscopic precession,* are put to work for pilots in three flight instruments: *attitude indicator, heading indicator,* and *turn indicator.* They're found on nearly all training aircraft, are indispensable for flight in instrument conditions, and, properly interpreted, can provide all the information you need relative to the performance of your airplane.

The three gyro instruments rely on continuous and consistent spinning of a high-mass rotor within the instrument case. Power for this rotation is furnished by an electric motor (usually mounted in and part of the rotor itself) or by a stream of air directed against "buckets" cut into the rim of the wheel. (Airplanes intended for instrument operations will have the power sources split, with a vacuum pump to spin the gyros in the attitude and heading indicators and electrical power to keep things going for the turn indicator. That way, when all is lost, all is not lost.)

KING OF THE INSTRUMENTS

The attitude indicator deserves royal classification, because it tells the pilot more than anything else on that panelful of instruments. At once, it provides information relative to both pitch and bank attitude, which in the conditions associated with normal flight maneuvers say a lot about what the airplane is doing in terms of climbing, descending, and turning.

Cut an attitude indicator apart and you'd find the gyro mounted horizontally in a set of gimbal rings, the whole ar-

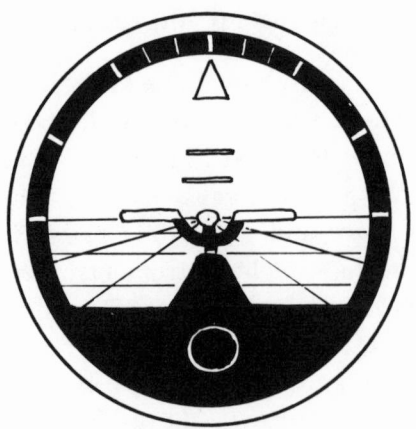

Figure 24

Attitude indicator

rangement set up to permit the rotor to remain stationary (i.e., horizontal) no matter what the airplane does. Out front, you see a representation of the horizon and a miniature airplane symbol, which gives the impression of observing movements in pitch and bank from somewhere behind the airplane.

Accurate, reliable, and just as sensitive as the pilot wishes it to be, the attitude indicator is an *instrument* instrument. Its greatest contribution to flight comes when you lose sight of the real horizon in clouds or haze, or a plastic hood when you practice in IFR conditions simulated by a plastic hood.

A SENSELESS COMPASS

All the problems of the magnetic compass—the swinging, deviating, oscillating false readings—are wiped out when you install a directional gyro. The DG is everything the magnetic compass isn't: It's stable, reliable, indicates a right turn when you turn right and vice versa, isn't affected at all by acceleration, and has its numbers arranged around a circular card so that you can see where you're going, directionwise.

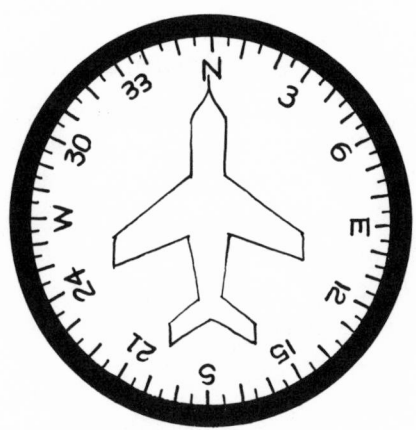

Figure 25

Directional gyro

Now the bad news: The directional gyro is a very dumb instrument. It doesn't sense anything, and provides valid information only when it is referenced to some known directional aid like . . . you guessed it, the magnetic compass.

Behind the card on the instrument panel is a gyro rotor mounted horizontally but arranged so that it stays put when the *heading* of the aircraft is changed; the force holding things steady is gyroscopic instead of magnetic. As you turn around the gyro, heading change is translated to a numbered card revolving under a pointer to show the heading of the aircraft. More often than not, the pointer is a symbolic airplane, which makes for instant interpretation by the pilot.

Once properly set, a brand-new DG should hold the correct heading for an extended period of time, but as the bearings and mechanism inside begin to wear, the heading indications will start to wander a bit. Should you blindly follow the DG as it creeps off the proper heading, you'll wind up flying in huge circles, wondering why you're over Pittsburgh instead of Paducah. The remedy is a simple one: Set the DG on a known direction before takeoff, and check it every fifteen minutes throughout the flight. What better-known direction than the concrete compasses at every airport? Set the DG on the runway heading just before you start rolling. You'll soon form the habit of taking a look at the magnetic compass every quarter hour or so, resetting the directional gyro when required. Enjoy the benefits of the directional gyro, but remember that it's only as good as the information it gets. Be certain that the wings are level and the airspeed constant when you take a reading from the magnetic compass.

WHICH WAY? HOW FAST? HOW GOOD?

Those are the questions answered by the turn-and-bank, turn-and-slip, needle-ball turn indicator, turn coordinator, or whatever you want to call it (them). Regardless of the name, there's an instrument in your panel that will tell you in which direction you are turning, at what rate, and whether you have coordinated your turning efforts. It's really two instruments, one to indicate direction and rate, the other quality.

Which way and how fast are measured by a gyro mounted in a significantly different manner than the other two gyro instruments just discussed. In the turn indicator, the rotor spins vertically and is free to move in only one axis; it can tilt, and that's all. Where the other two gyro instruments utilize the gyroscopic principle of rigidity in space, the turn indicator depends on gyroscopic precession to let you know what's happening.

Put simply, when you try to tilt a spinning body, it turns; attempt to turn it and it tilts. Since this gyro can *only* tilt, it will do just that whenever the heading of the aircraft is changed, thereby trying to turn the gyro. The tilting is translated to the pointer (sometimes a symbolic airplane) on the face of the instrument, and shows you by deflection left or right which way you're turning. The displacement of the indi-

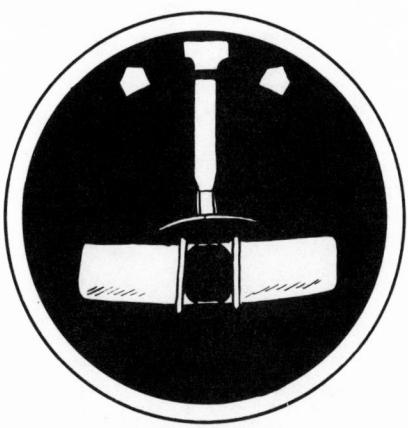

Figure 26

Turn-and-bank indicator

cator tells you *how fast*. A VFR pilot shouldn't have any doubt about which way he's turning, and the rate of turn is academic when you can see outside. Get inside a cloud or under a hood, however, and these indications become significant.

The other part of the turn indicator is simplicity itself: nothing more than a scaled-down carpenter's level, which lets you know when the force of gravity is acting straight down or pulling off to one side. A small ball in a slightly bent glass tube filled with liquid is free to move left or right in response to gravity (or centrifugal force in a turn) in the same way you and everything else in the airplane respond to gravity (or centrifugal force). If you enter, maintain, and roll out of a turn properly, the ball will stay put in the center of the tube, and you're to be congratulated on your fine coordination of bank angle and rate of turn. Get into a "slip" (the airplane is banked too much and falls toward the low wing) and you'll feel yourself pressed against the low side of the airplane; take a look at the ball and you'll see it's out of kilter on the same side. During VFR training, it's better to develop a sense of coordination through the seat of your pants, using the ball to confirm your suspicions.

If nothing else, the turn indicator is simple and super-reliable; it is error-free so long as the power source is sufficient to keep the rotor spinning at the proper speed. Even if you somehow get upside down, the turn indicator does its things, always telling you which way, how fast, how good.

AUTOMATIC PILOTS

Our society has a longstanding record of adapting machines to do most of our work—using our heads instead of our backs, as it were. Why shift gears when an automatic transmission does the job so much better and easier? Why wrestle a steering wheel when you can command the genie of hydraulic pressure to turn the wheels? And eventually we're sure to see the introduction of personal ground transportation systems that will require little more than coded instructions to automatically follow a buried cable or a radio signal from here to there.

In consideration of the fact that airplanes and automobiles have grown up more or less together, there appears to be a wide gulf between the two in terms of making things easier for the operator. We don't have to pull the prop through to start an aircraft engine anymore, but most of the actual work of flight is still accomplished with muscle. That it doesn't take much muscle is a tribute to good aerodynamic design, not innovations in aircraft-control systems.

As far back as 1914, aero-engineering pioneers were grinding away at the problems of automatic flight, and were being ground down by the old bugaboo of weight compromise; the gyroscopes and other gear which seemed necessary for success were just too heavy for the airplanes of the day. Autopilots would have to wait until either the aircraft grew in size or the flight-control industry was able to produce smaller, lighter components. As it turned out, both of these came to pass—today's big airplanes could carry even the heaviest of yesteryear's autopilots with no problem, and space-age miniaturization has shrunk the electronic components of autopilots to a size and weight compatible with nearly any airplane in the fleet.

The basic principles of automatic flight operation are sensing, computing, and controlling—a progression of events not a whit different from the processes you go through to fly an airplane manually. For any given attitude or other condition of flight, the autopilot must somehow acquire information (sensing, usually done by means of gyros), interpret that information, and decide what needs to be done (computing, the job of the small analog computer that is the heart of any autopilot), and finally exerting pressures where they are needed to maintain or correct the condition (controlling, through electric or hydraulic servo motors attached to the flight controls). The information delivered to the computer can be augmented by navigational radio signals and air-pressure data in the more sophisticated models, so that the autopilot becomes a first-rate navigator and a precise maintainer of altitude and airspeed.

In its simplest form, an autopilot is known as a *wing leveler*, and it does exactly what the name implies. By sensing the aircraft's banking attitude through a specially wired gyroscope, this type of autopilot applies whatever aileron pressure is necessary to keep the wings parallel to the horizon. When it's time to turn, the human pilot must either override the leveling pressures or press a button on the control wheel to temporarily disconnect the autopilot. When the turn is completed, release of the button or relaxation of aileron pressure allows the wing leveler to take over once again. In consideration of long-term flight with no heading change, some of these simple autopilots incorporate a trim circuit, which can be adjusted to correct any tendency for the airplane to creep right or left.

The next level of autopilot sophistication is achieved with the addition of a *turn control,* which merely includes another servo motor for the rudder. In this installation, the human pilot can command coordinated turns by moving a knob or switch. Add a control circuit for the elevators, and you've a complete unit—a three-axis autopilot—which is really "power steering," in a sense. Quite simply, instead of supplying arm and foot power to deflect the control surfaces, you direct a group of small electric motors (large airplanes generally employ electro-hydraulic systems) to do your bidding. It's still necessary for the human pilot to make the decisions, but when the airplane is pointed in the proper direction at the right altitude and air-speed, the man can lean back and relax, with more time available to concentrate on what he's going to do next.

An autopilot might be further augmented with the capability of interpreting certain radio-navigation signals. For example, when programmed to intercept a VOR radial, such an autopilot will turn the airplane toward the displaced left-right needle and roll out on course when the needle centers. With the left-right needle wired so that an electrical signal is generated whenever it's off center, any displacement results in a turn command from the computer; when the needle's centered, the autopilot holds whatever heading is required to keep it centered, and wind drift is taken care of automatically.

From this point, it's a relatively simple matter to tie an adequate autopilot system to localizers and glide slopes for completely coupled instrument approaches. All the pilot needs to do is adjust the power setting to control airspeed (autopilots generally keep you on the glide slope by changes in pitch), sit back, and watch for the runway.

The addition of air-pressure data enables the autopilot to hold a desired altitude or climb/descend to a preselected level. This is usually done by incorporating a separate air-pressure bellows system in the autopilot black box. When engaged, the altitude-hold function sends minute electrical signals to the computer whenever air pressure changes the slightest bit, and the elevator servos are operated to bring the airplane back to the original altitude. With the additional feature of altitude preselect, the human pilot can dial in the altitude he wants, and the computer commands a climb or descent as appropriate and levels the airplane at the proper time.

These "big airplane" features—capturing and maintaining an electronic course, flying an instrument-approach procedure, climbing or descending to a specific altitude—require a modest amount of computer programming from the pilot himself. You'll find a variety of presentations, but all such installations include a panel with push buttons or switches or levers to select the type of operation desired. Even the modest autopilot systems may have additional features and benefits to help make your flying more pleasant and safe, so consult the handbook to be sure you're setting it up properly.

And don't let any of the autopilots-are-for-sissies old-timers sell you a bill of goods. There's absolutely nothing wrong with using an autopilot to help you with the chores of flying, especially when you're flying by yourself. Who can argue against letting the machine take care of itself while you spend more of your time looking outside for other airplanes? Who can say that it's not a far better thing to let the autopilot handle the busy work on a long trip so that you'll be refreshed and sharp when landing time comes?

There is one ominous note, however: He who lets the autopilot do all the work (you can't legally allow the black boxes to execute takeoffs and landings . . . not yet) gradually loses the ability to do it well himself. When you fly regularly in an autopilot-equipped airplane, make it a point to fly by hand often enough to maintain the basic skills. Even the best autopilots have been known to roll over and die now and then.

Flying When You Can't See Outside

7

Here's one right out of the twilight zone: Having just learned to drive, and before leaving on your first long trip, you are instructed that, no matter what happens, you must continue straight ahead at 60 miles per hour and everything will work out. For reasons you're not prepared to understand, you accept this condition and start on your way. Next thing you know, you're rolling along in the middle of the night, full moon rising straight ahead in a cloudless sky, and suddenly the stripes on the pavement disappear—no center line, no edge-of-the-highway stripes, nothing—and, to compound the problem, the headlights go out.

Your first impulse is to stop, but your instructions (which you feel strangely compelled to follow) were very explicit: Continue straight ahead at 60 miles per hour. With nothing to lose by trying, you put your faith in what you've been told and press on into the night. After a while, with the moon for a guide and the speedometer needle glued to 60, the lights come on, the highway stripes reappear . . . and you wake up from a very bad dream.

For a number of years, a demonstration of one's ability to control an airplane without the benefit of outside visual references has been a condition of pilot certification. An appalling percentage of general aviation accidents occur when pilots with no instrument capability fly into low-visibility conditions. With just a smattering of instrument training, however, a pilot can keep his aircraft right side up and fly out of danger. The big difference between this and the automotive nightmare is, of course, the third dimension, altitude.

Before we launch this discussion of how to use those flight instruments, be aware that there's also a tremendous mental adjustment which must take place whenever visual clues are lost. You must put absolute faith in the instruments, because the normal inputs of feel and sound are capable of making you think the situation is exactly the opposite of what's really going on. It's a very difficult thing to do—even high-time pilots often find their senses in disagreement with the instruments—but the *only* way to fly out of trouble is to disregard what seems to be happening and substitute the instrument indications for what you were able to see outside before the headlights went out. With a little bit of luck (and probably a lot of help from someone in Air Traffic Control), you'll soon be flying in the sunshine again, or at least in conditions that let you see where you're going.

HOW FIRM A FOUNDATION?

Unless it suffered from serious steering-gear problems, your blacked-out Nightmare Eight would probably have continued straight ahead with little or no help from you, and so long as you kept the throttle in one position, the speed would likely have remained right on 60. If the roadway ran up a hill, you'd have been obliged to add some power to keep the speedometer steady, and vice versa going down. You must believe that your

---

airplane—*any* airplane—will behave very similarly. The secret is *trim*.

Let's say you're flying straight and level at a normal cruise airspeed with the airplane trimmed for hands-off flight. (Is there any other way?) For the sake of experimentation, reduce the power setting—perhaps two inches of manifold pressure, or a couple of hundred rpm—and, with no pilot input, the nose will drop just enough to maintain the airspeed for which you had trimmed. You'll start losing altitude, of course, but that's the only way the airplane can make up for the subtraction of thrust—remember, the trim tab was set for cruise airspeed, and the elevators change the pitch attitude whenever that balance of aerodynamic forces is upset.

When the airplane has descended far enough in this reduced-power condition, return the throttle to its former setting. Lo and behold, the nose will rise to the original cruise attitude, and the airspeed will stick on the original number. Want to climb? Merely add power beyond the cruise setting; up goes the nose in order to maintain the trimmed airspeed, and the airplane climbs. You have proved that trim controls a constant airspeed, and that the throttle reigns supreme over altitude.

When you change the power setting to climb or descend, or stop doing either of those, the change in pitch attitude varies considerably from one aircraft design to another. Some instantly assume and hang on to the new attitude; others oscillate somewhat until the pitch attitude stabilizes. How a plane reacts is largely a function of the number of design and manufacturing dollars spent on pitch stability. After you get accustomed to the way your airplane reacts, you can help it a bit with a touch of pressure on the control column or stick to minimize the excursions in pitch.

One of the conditions of the foregoing experiment was "no pilot input" except the power changes. You could get away with that if your airplane didn't have a propeller, but the changing forces of torque and asymmetrical thrust cause heading changes unless you put a stop to it. In general, whenever you add power the nose will swing to the left; when you reduce power, the nose will swing to the right (all this at a constant airspeed, of course). It's your job as manager of this system to prevent those heading changes, assuming that the objective is straight flight. So rudder trim comes into the picture.

You'll recall that all propeller-driven aircraft suffer somewhat from asymmetrical thrust, even at cruise, which leads to the vertical stabilizer's being offset just enough to take care of the problem at a predetermined airspeed. Whenever you are flying slower than the stabilizer-offset speed, it will be necessary to supply additional right rudder pressure (or trim) to keep the nose from swinging to the left. Should the airspeed build up beyond the designer's zero-trim number, the nose will slew off to the right unless you apply whatever rudder pressure (or trim) is necessary to prevent it.

Assuming now that you have the airplane *completely* trimmed (that will mean just what it says if your airplane is equipped with a rudder trim tab, or whatever rudder pressure is required if it's an elevator-trim-only airplane), run through that experiment one more time, supplying no elevator pressures but whatever rudder pressure is necessary to keep the nose from moving to the left or right. You'll find that as power is reduced for descent, a little left rudder pressure must be fed in to keep the heading from changing; the opposite occurs when you add power to either level off or start a climb. How much rudder pressure? That's easy—whatever is required to keep the nose on a point, which might be a lake or a highway or a distant building when you can see outside, and which *will* be a specific number on the heading indicator when you've flown yourself into a cloud.

FOUNDATIONS ARE MEANT TO
BE BUILT UPON

When you've convinced yourself that the airplane will fly level, descend, and climb pretty much by itself, with your input limited to whatever rudder pressure you need to maintain heading and occasional touches of elevator pressure to stop a pitch oscillation before it starts, you're ready to get down to the brass tacks of controlled, predictable flying when you can't see outside.

The procedures that you're going to build on that solid foundation of one properly trimmed airspeed are the procedures to be used when maneuvering is required to get you back into VFR conditions; if that fails, they're the lifesavers that will get you back onto the ground in instrument conditions. Emergency? You bet. And when the chips are down, it's important that you deal with familiar attitudes, power settings, and aircraft configurations; consider them working procedures, to be used with confidence when anything other than straight and level flight is required. But in the more likely situation, where you fly suddenly into less-than-visual conditions at normal cruise airspeed and a safe altitude, common sense dictates that you leave everything right where it is—assuming proper trim, of course—and continue in the direction of better weather with all your mental powers and manipulative skills devoted to solving your problem. The airplane will fly itself.

The simplest extraction procedure would consist of recognizing a low-visibility problem, contacting someone in ATC, and following their instructions regarding which direction to fly to get out of trouble; that's the situation best handled by continuing at whatever speed you're trimmed for when it happens. The most complex problem you'll ever face as a non-instrument-rated pilot would be that bona fide emergency in which you have allowed yourself to be completely surrounded by inclement weather and have no choice but to fly all the way to the nearest airport and land under instrument conditions. With numerous turns, perhaps some climbs, and at least one descent (How else can you get onto the ground?), you need some working numbers, built on the foundation of one previously proven trimmed airspeed.

Your personal survival airspeed should be a compromise. You should consider an adequate margin above the stall speed, fast enough to provide crisp control responses and prevent over-oscillating, but not so fast that events occur at a pace you can't handle. If your airplane goes faster than your mind, you'll be in worse trouble. For most of the single-engine aircraft in the general aviation fleet, an airspeed of 100 to 120 miles per hour will feel about right; heavier singles and twin-engine aircraft will handle better at higher airspeeds.

Of less importance than the airspeed itself is knowing what conditions produce that speed. You need to know a pitch attitude and a power setting which will maintain the airplane in level flight at, for example, 100 miles per hour. (The numbers used from here on are for illustration only; they may not apply to the airplane you're flying, but they should provide some rough relationships for your experimenting.) When the airspeed is steady and the altitude remains constant, note the pitch attitude. Adjust the attitude indicator until the little airplane is resting precisely on the horizon, take a mental photograph of the attitude indicator if it's not adjustable, or hold the altimeter motionless with elevator pressure if you've no attitude indicator at all, and trim until there's no pressure on the wheel or stick. At the same time, read the power instrument. With a fixed-pitch propeller, the tachometer might indicate 2000 rpm; manifold pressure with a constant-speed prop might be at 15 inches. A little bit either way in airspeed or power setting is relatively unimportant—the key point is to arrive at a target power setting and pitch attitude.

Now prove your point. Change only the power setting and fly around the sky for a couple of minutes, until the airspeed and attitude are well displaced. When everything is out of kilter, return the throttle to the target setting, level the wings, put the nose in 100-mile-per-hour attitude (stop the altimeter with elevator pressure if no attitude indicator is installed), and sooner or later the airplane will settle down in level flight at 100 miles per hour. It works!

Take your hands off the wheel, use rudder pressure to hold the nose absolutely motionless on the heading indicator, and slowly add power up to 20 inches (you might as well go to full throttle with a fixed-pitch prop). The nose will pitch up a bit as you open the throttle, but in a very short time it will establish a new 100-mile-per-hour attitude, and the airplane will be climbing—at about 500 feet per minute—in response to the added thrust. About 20 feet from the desired altitude, just as smoothly bring the power back to the level-flight setting—some left rudder pressure will be required to keep the nose from swinging—and your flying machine will reestablish itself in level flight at the trim airspeed.

A descent is just as impressive, and results when you reduce power by about 5 inches or 500 rpm. The nose will pitch down just enough to maintain the airspeed at 100, the thrust you took away will be replaced by the force of gravity, and the airplane will descend at approximately 500 feet per minute. Level off by replacing the power you removed, but because of the combined forces of gravity and inertia, lead the desired altitude by 50 feet. Like magic, the nose pitches upward to maintain the trimmed airspeed and you're in level flight once again. All of this, by the way, should have happened "hands off"; whatever rudder pressure keeps the nose straight will also keep the wings level. If nothing else, you've proved that the airplane gets along quite well by itself, that it won't fall out of the sky when you let go of the wheel. That should make all your flying a bit more effortless from now on!

YOU CAN'T ALWAYS GET THERE FROM
HERE IN A STRAIGHT LINE

Just as you worked out power settings to produce a moderate airspeed in level flight and comfortable rates of climb and descent, any turns on instruments should be gentle ones, enough to change the heading at a reasonable rate, but not so much that the airplane starts to get out of hand. Try 10 degrees of bank for your first turns "on the gauges." The heading change will be slow but steady—after all, in a situation like this you're not in a hurry—and, perhaps more important, the redistribution of vertical lift in a 10-degree bank is so slight that very little back pressure is required to maintain altitude. When the altimeter starts to slip downward (and it will, sooner or later), apply just a touch of back pressure to the wheel to hold the altitude pointer where you want it. The pressure is so slight that you can supply it quite literally with a fingertip.

On most light aircraft, the ailerons are the most responsive controls, and stability around the longitudinal axis—the roll axis—is usually less than that in either pitch or yaw. Of all the control inputs available to the pilot, the one that upsets things most rapidly and to the greatest degree is roll. With this in mind, all instrument turns should be entered, maintained, and terminated with rudder alone. In a situation where you'd like to keep things on as even a keel as possible, keep your hands away from those controls that have a high probability of upsetting the apple cart—using the ailerons will almost always result in overcontrolling, which leads to over-*un*controlling.

Fold your arms across your chest, hold hands with your co-pilot, sit on your thumbs if you must, but when you get ready to turn, forget the ailerons and apply very gentle pressure to the rudder pedal. As the nose yaws—say, to the left—the right wing will begin to move through the air faster than the left one (it *must*, being on the outside of a circle now), and of course develops a bit more lift from the added airspeed. Slowly,

---

gently, the right wing will lift to provide precisely the right amount of bank for the rate of turn you've induced. When 10 degrees of bank is established, relax the left rudder pressure. Throughout the turn you can make small adjustments with the rudders to maintain the bank angle at 10 degrees. (If you've no attitude indicator to show you how much bank, press on the appropriate rudder until the turn needle moves half its width from the center index; the resultant bank angle will be very close to 10 degrees.) Rollout should be commenced with slight rudder pressure when the heading indicator is 10 degrees from the desired number, and the pressure on the pedal should be just as firm and smooth as it was when you started the turn. Once on the desired new heading, keep it there with rudder pressure until the airspeed settles down once again. The price of a level turn during which no power was added is a little bit of airspeed, but have patience, everything will return to the original numbers sooner or later.

With practice, you may want to increase the rate of turn by steepening the bank—no problem, so long as you realize that additional back pressure will be required to hold altitude. Also with practice, there's no reason why you shouldn't become proficient at descending and climbing turns—it's just a matter of doing two things at once, and, if nothing else, such practice can become a challenge to your flying skills.

Instrument flying, even at the very lowest proficiency level, is a rather specific, demanding skill. If it isn't practiced once in a while, it's soon forgotten and is largely irretrievable when needed most. Once learned, the ability to keep an airplane right side up using only what you see on the instrument panel must be practiced occasionally just to keep rust out of the hinges. When instrument flight is the difference between success and failure, the state of mind that usually exists is not conducive to calling up long-unused skills.

Weather

8

If all the world's pilots nominated the condition that most often affects their flying activity, the chances are very good that weather would be the winner. While there are as many reasons for not flying as there are pilots—money, time, maintenance, sickness—when it comes right down to the wire, when an airplane is available and the pilot is ready, every single flight must contend with weather, for we live at the bottom of an ocean of air and are therefore subject to whatever changes take place in the lower levels of that vast blanket of gases that covers the earth. Aviators must attach special significance and respect to micro-meteorology (the weather you can see from the ramp at the airport) as well as the weather along your proposed route of flight, because modern aircraft *create* changes by carrying you rapidly into and through areas where the weather may be completely different from the conditions at takeoff.

"Weather" is a comprehensive word that describes the state of the atmosphere in terms of humidity, temperature, pressure, visibility, wind, clouds, and precipitation. Weather, like most complex subject areas, can best be understood by going straight to the source—by studying the properties of air and the changes brought about by heating it, cooling it, adding moisture or taking it away, and the changes caused by the very movement of air across the face of the earth. Air behaves somewhat predictably—otherwise forecasters would be regulars at the unemployment office—but you should always allow for those unexplainable quirks of nature that can make fools of even the finest weathermen.

HEATING AND COOLING

The sun makes it all happen; without the radiant heat from our mother star, the Earth would be seasonless, windless, and, of course, lifeless. Solar radiation is the source of local daily temperature changes as well as the large-scale heating that makes the difference between tropical and polar climates.

When air is *heated*, several things can happen: The air expands, the vertical currents thus produced clear the air by carrying upward whatever pollutants exist, and visible moisture—clouds, haze, fog—reverts to invisible water vapor when sufficient heat energy is added.

When air is *cooled*, the opposite reactions generally take place. The air contracts and tends to settle, carrying with it to the lower levels whatever restrictions to visibility might exist, and trapping man-made pollutants near the surface. If there is sufficient water vapor in the air, the cooling process may cause condensation and a resultant reduction in visibility.

In addition to solar radiation, there are several other ways in which the temperature of the air can be changed. Rain often drives the mercury downward because of evaporative cooling, air moving over snow-covered ground or frozen lakes tends to cool by contact, and the radiation of heat by deserts and light-colored areas in strong sunlight tends to raise the temperature of the air directly above. Vertical movement of the air can also change the temperature without the addition or subtraction of heat; such is the case when the wind happens to be blowing up a mountain slope—the air expands as it rises, and the temperature drops because the same amount of heat energy is now distributed throughout a much larger volume of air. The reverse takes place when air moves downhill—it warms because of compression.

Under standard-day conditions, air temperature drops 2 degrees Celsius for each 1000 feet of altitude gained, but the gods of weather never heard of a standard day, and in the real world the temperature lapse rate is frequently different from book value. How much it is different, and whether it is higher or lower, determines the *stability* of the atmosphere. A particular quantity of air that cools at the same rate as the surrounding air tends to arrive at any given altitude at the same temperature, and therefore the same density, as the air through which it rose. This is *neutral stability*, with the lifted air neither rising nor falling.

On the other hand, a parcel of air that cools more rapidly than the air through which it's rising always has a lower temperature than the surrounding air, is therefore more dense, and exhibits a tendency to sink through the atmosphere—a stable condition, in which the air returns to its original level in the atmosphere when the lifting force is removed. Instability occurs when air cools less rapidly than surrounding air, is always warmer and less dense, and tends to keep right on rising even though the lifting force ceases to operate.

In general, you should expect instability when flying through warm, moist air (water vapor tends to promote instability, because a great deal of heat is soaked up in the process of evaporation, thereby slowing the lapse rate), especially if it is being heated from below—the usual summertime situation in all but the very dry areas of the country. Look for stable conditions in the winter, or whenever air is moving over colder surfaces. The

vertical currents that are part and parcel of unstable air develop cumulus clouds, while stable air produces flat, layered
stratus clouds. Flight through stable air is smooth, but expect
the number and severity of the bumps to increase as the stability of the air decreases.

ADDITION AND REMOVAL OF MOISTURE

Any pilot who operates on the downwind side of a large body
of water should be aware of evaporation and its role in the
weather he'll experience. Air moving across open water soaks
up some of the water vapor produced by evaporation; when
water and air are both warm, the amount of moisture carried
inland increases and affects the weather remarkably. Given the
instability that the added moisture promotes, any lifting action
is likely to trigger towering cumulus clouds, but if the moisture-
laden air happens to move downwind over a very cold land
mass, expect low stratus clouds.

The other major moisture-adder is precipitation. More so
with rain than with other types of falling water (sleet, hail,
snow), some of the moisture evaporates into the air and increases the humidity. If it's raining on your parade and the air
is cool enough, look for low clouds to form when the rain puts
more moisture into the air than it can possibly hold in a vaporous state. (At the other extreme, in dry parts of the country
you will often notice precipitation from high clouds that
evaporates completely before it reaches the ground. The air is
so hot and dry that the water changes from liquid to vapor
very quickly.)

While precipitation increases the water content of the air
through which it falls, the air that produced the rain to begin
with must necessarily become drier. The removal of moisture
from a given parcel of air increases its stability, makes it less
likely to produce clouds, and in general causes an improvement
in the weather. Heating appears to remove visible moisture

from the air (as on a gray morning when the fog burns off halfway to lunchtime), but it's really only driving the liquid water back into a vaporous state and distributing it throughout a greater volume of air. If that air sticks around until the end of the day, the fog will come sneaking back just as soon as the temperature drops.

AIR-MASS WEATHER

No one knows what really goes on in the atmosphere, and probably no one ever will, because we can't see the air, only the results of its changes and movements. But meteorologists are fairly confident of the existence of huge bubbles of air resting on the surface of the Earth and moving about under the influence of solar heating. When such a parcel of air can be isolated from the surrounding atmosphere on the basis of temperature, moisture content, stability, and pressure, it can be identified as a specific *air mass*. Knowledge of its general properties, where it came from, and the terrain over which it's moving can provide a forecast—admittedly a very general one —of the weather conditions that particular air mass should produce.

Source regions—those areas of the Earth's surface from which air masses come—vary in the quantity of heat and moisture they contribute to these atmospheric bubbles, but on a wide-ranging, relative scale they can be classified as either maritime or continental (wet or dry), tropical or polar (warm or cold). In combination, these qualities speak loudly for the subsequent behavior of an air mass. Dry, cold air (labeled *continental polar*, after its source region) will probably produce smooth, cloudless skies, while a *maritime tropical* air mass with its abundance of moisture and attendant instability will no doubt develop cumulus clouds and make for bumpy flying.

Given the general movement of an air mass (west to east in

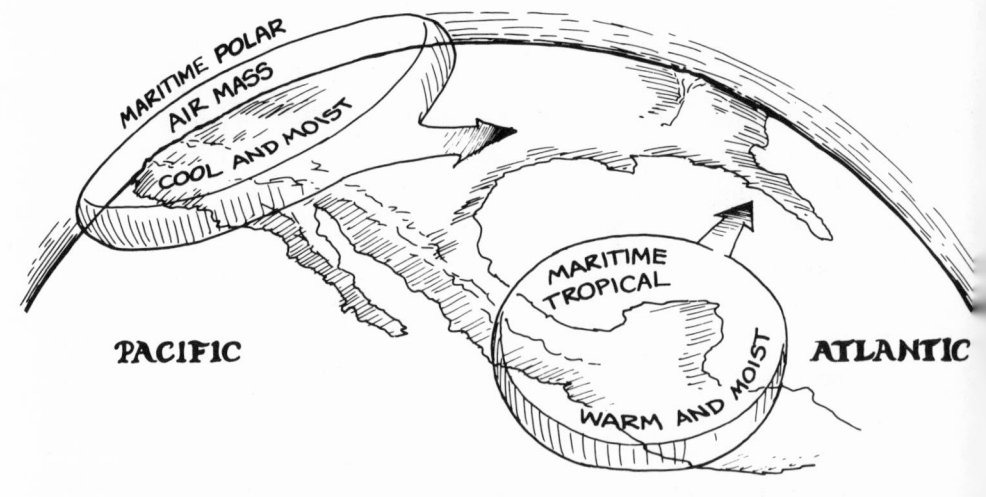

Figure 27

Two types of air masses common to the continental
United States, and their source regions

the temperate zone of the Northern Hemisphere, because of
the prevailing westerly winds) and knowledge of the properties imparted by the source region, you can predict air-mass
behavior as it is modified by the surface over which it moves.
One of the most dramatic examples of such modification occurs
in the southeastern United States in wintertime, when warm,
moist Gulf of Mexico air (a maritime tropical air mass) flows
over the Southern states. When the temperature is lowered by
contact with the relatively cold ground, condensation (clouds)
and settling take place, and all of Dixie may experience instrument conditions for days on end. At the opposite extreme, the
same warm, moist Gulf air might move directly eastward over
the sun-warmed Florida peninsula and be modified by rapid
heating from below, with much lifting action, cumulus clouds,
and daily thundershowers. You can almost always spot the
towering clouds that build up above tropical islands long before the land itself becomes visible.

Yet another type of modification takes place when an air

mass pushes across mountain slopes. The temperature change is caused by the normal expansion and cooling due to altitude, and if the air is moist enough, clouds may form at the higher elevations. The air mass may be dried considerably as a result of precipitation from the clouds, and when the drier air starts down the other side of the mountain, it is compressed by the change in elevation and may wind up very much warmer than normal. This phenomenon is responsible in part for the cherry blossoms in Washington, D.C., several weeks before spring arrives on the other side of the mountains.

An example of terrain modification that has changed the face of the land is found in the Pacific Northwest, where cool, moist air from the Pacific Ocean moves eastward and is lifted by the Coastal Range. The nearly continuous precipitation on the ocean side has created and maintained rain forests, but after the rest of the moisture is squeezed out by the high Cascades, the warm and dry air descends into the desertlike terrain of eastern Oregon and Washington.

CLOUDS

The real nemesis of VFR-limited pilots, clouds can be highly localized (fair-weather cumulus sailing across an otherwise clear sky) or a general condition, with fog and haze covering thousands of square miles. No matter what the type, clouds are nothing more than moisture vapor that has become visible because of condensation, usually the result of lowered temperature.

All air contains some moisture, visible or not. One of the measures of moisture is *relative humidity*—the amount of water vapor in the air, expressed as a percentage of what that air will hold as invisible moisture vapor. Relative humidity is missing from the weather information supplied to pilots; we're more interested in the air's water content in relationship to its temperature, because condensation will occur in any parcel of air when the temperature reaches a certain value.

As the temperature drops, the ability of the air to hold water as an invisible vapor decreases. When the air is very moist (high relative humidity) the temperature needs to be lowered only a small amount for condensation to occur, and clouds form readily. Very dry air must be cooled a great deal before any of the water vapor becomes apparent to the eye; but, regardless of the amount of moisture, the temperature at which condensation takes place is known as the *dew point*, and information about it is important to pilots and is furnished as a part of all hourly weather observations. When the air temperature reaches the dew point, some sort of condensation takes place, and the moisture becomes visible. For example, in the micro-atmosphere around your gin-and-tonic, the temperature of a hot summer day is lowered to the dew point by the iced drink, and water condenses on the outside of the glass. The same thing happens in the atmosphere—water vapor condenses on particles of dust and debris in the air—and the result is a cloud.

Any process that causes the temperature to drop to the dew point can be responsible for cloud formation: radiation cooling after sunset, when the air close to the earth is cooled by contact with the cold ground, or perhaps an air mass being lifted over a mountain range and cooling as it expands. In either case, if the air is cooled to its dew point, condensation occurs. (By the same token, when the temperature somehow gets back above the dew point, the clouds begin to disappear, which explains why early-morning fog goes away as the sun warms the air.)

If clouds form whenever the temperature drops to the dew point, it should be possible to turn things around and build a cloud by raising the dew point to the temperature—and that happens frequently when rain falls through cool air, evaporating as it descends. When cool air passes over water, the additional moisture is often enough to trigger condensation—sea fog—close to the surface.

There are two cloud forms—cumulus and stratus—that are founded primarily in the stability of the air that produces them. (Cirrus clouds, those fine, wispy brush strokes against the blue sky, are of no concern to low-altitude pilots except as precursors of approaching weather. Cirrus clouds are always made up of ice crystals, and are found only at very high altitudes.) Cumulus cloud formations are heaped-up masses, cauliflowerlike on top, while stratus clouds spread out in relatively uniform layers. Of course, in a well-developed weather system there may be infinite combinations of these two basic types.

The presence of cumulus clouds always implies vertical currents in the air. They are usually formed by heating from below, and are associated with unstable atmospheric conditions; air that does not cool as rapidly as the surrounding air is always somewhat warmer, less dense and buoyant, and tends to continue upward when any sort of lifting action is introduced. These vertical currents are very graphically portrayed by the "chimneys" of cumulus clouds in the summer sky—local solar heating creates updrafts, which lift warm, moist air to an altitude at which the dew point is attained, and a cloud forms. Every cumulus cloud is the visible cap of a column of rising air . . . the soaring pilot's delight, because it lets him know exactly where lift is located. As often as not, the bases of these summertime clouds are very flat and uniform, indicating clearly the altitude at which the dew point and the temperature have come together.

The distance between the ground and the cloud bases is a good indicator of the temperature–dew point spread. When the bases are very high, the spread is very wide, and vice versa. Conversely, when you know that there are a lot of degrees

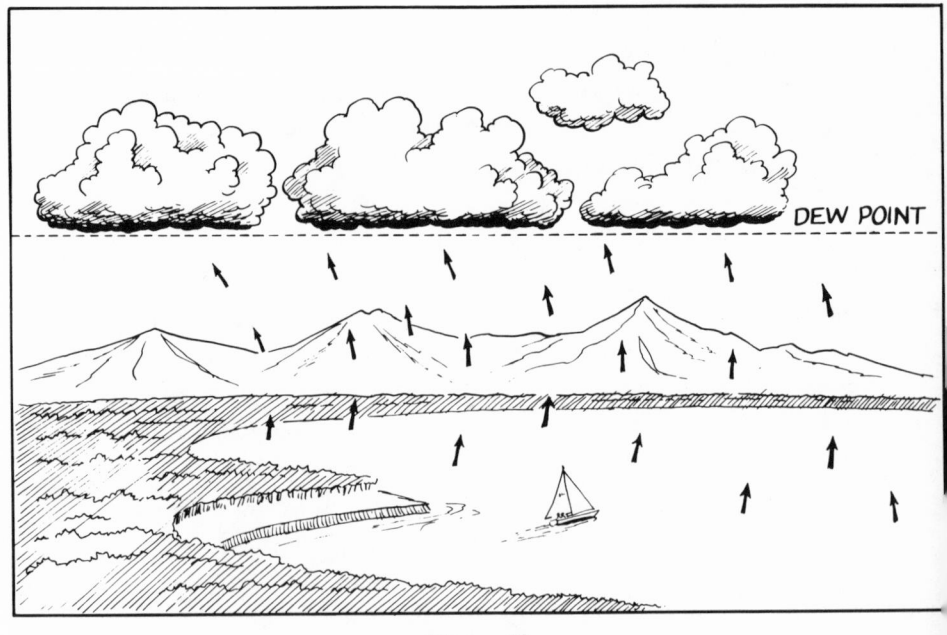

Figure 28

Cloud bases indicate the level at which the air temperature
has cooled to the dew point

between temperature and dew point, you can expect a lot of
clear air between the cloud bases and the ground.

Cumulus clouds also happen when unstable air is pushed
aloft by rising terrain. It matters not at all what the lifting
action is; when unstable air reaches the altitude at which its
temperature is nearly the same as the dew point, a cumulus
cloud will begin to take shape.

Almost completely opposite in nature, the stratus cloud
shares only the principle of formation with its cumulus counterpart; when stratus clouds are evident, you can count on a
lack of vertical movement in the atmosphere. Stratus clouds
are usually formed as a result of cooling from below, and are
associated with stable atmospheres; air that cools at the same
rate or perhaps even more rapidly than the surrounding air has
no tendency to continue rising, and forms stratified clouds that

seem content to stay put. The gray, flat skies of winter provide a good example, and when a warm southerly wind moves moist air over snow-covered ground, the dew point is reached over a very wide horizontal area.

Fog is *always* a stratus cloud. It's called "fog" only because of the proximity to the surface. Since there is little or no vertical movement within stratus clouds, they have characteristic table-flat bases and tops, and flying near and within is usually as smooth as glass. Stratified clouds will also form when the dew point is reached by virtue of a stable air mass crawling slowly up rising terrain. If the movement is slow enough, ridge tops will be completely obscured by cloud while the valleys remain quite clear.

As a general observation, the conditions that produce cumulus clouds are more favorable for VFR-limited pilots than those situations that result in stratus formations. The vertical currents associated with cumulus clouds usually carry most of the atmospheric debris (haze, smog, smoke, smaze, and all the combinations thereof) upward, distributing it over a wider and higher area. When there's little vertical movement, *à la* stratus, the restrictions to visibility are contained near the surface and probably will spread out over a larger area for a longer period of time.

Your knowledge of air-mass source regions and movement plus the principle of atmospheric stability and cloud formation should enable you to make at least a good guess about the cloud types and visibility to be expected from an approaching weather system. Since the net movement of air masses across the United States is from west to east, a cold and moist air mass will probably generate vertical currents and cumulus clouds if warmer areas of land or water lie in its path; a warm, moist air mass will no doubt build up layers of stratus clouds if and when it crosses a colder surface. A warm, dry air mass such as those generated in our own southwestern states is joy unlimited

—no clouds, clear skies, and it can move almost anywhere and still produce nothing but beautiful flying weather.

PRECIPITATION

Just because a cloud forms in the sky doesn't mean that there will be precipitation, but there can be no rain or snow or sleet or hail without a cloud to start the process. The cloud-building requirements—moisture and a small temperature–dew spread —must precede precipitation of any kind. No matter in what form it finally arrives on the ground, precipitation begins as microscopic water droplets that form a cloud. If the droplets get no bigger, the cloud won't precipitate, but introduce some movement inside the cloud and the drops of water collide, each one adding to the weight of the larger droplet, until at last it's too heavy to float and falls out of the cloud as rain. When the air is very cold, water vapor actually condenses in solid form and becomes snow; rain that freezes on the way down arrives as ice pellets; and droplets that bounce around inside a freezing cumulus cloud add layer upon layer of ice until they fall out as hail.

The type of cloud in which condensation occurs usually determines the type of precipitation. Cumulus clouds, in which there is considerable vertical movement and therefore considerable collision of water droplets, can be counted on to produce large-drop showers—the kind that really splatter on the windshield. Conversely, stratus clouds and their characteristic lack of intra-cloud movement produce misty, drizzly rain with very small drops.

You can also turn this principle around and get some idea of the type of cloud that's ruining your flying day by observing the size and intensity of the precipitation. Conditions that produce *drizzle* have therefore produced stratus clouds; you can infer a widespread weather system that will likely be a long time clearing out. Heavy showers with big drops (goose-drowners and gully-washers) indicate the presence of cumulus build-ups, and probably a short-duration, narrow band of weather.

WIND

If it were somehow possible to bring the Earth's rotation to a halt and shut off the sun for a while, the atmosphere would settle down into an equal-depth layer of air. But with the sun switch thrown, the atmosphere would soon develop hot spots, causing the air to expand and rise. Since the atmosphere is fluid, more air would move in to fill the void. That movement is *wind*. A forest fire makes a good, if extreme, example of this process.

The patterns of atmospheric pressure created by uneven solar heating are constantly shifting, and the Earth's winds blow in response to these changes—always moving air from high pressure to low, and with greater strength where the pressure differences are large.

The all-time champion hot-spot is the equator, where the sun's rays strike the Earth nearly head-on. The air thus heated rises rapidly and produces a low-pressure area, which is replenished by surface air. When the rising air reaches the upper limit of the atmosphere, it divides, one stream proceeding north, the other south. (Once again, you need to be concerned only about the Northern Hemisphere. Everything is reversed on the other side of the equator.)

In the absence of any other influences, the equatorial air would no doubt proceed along the "ceiling" of the atmosphere all the way to the North Pole, then flow southward in a steady stream to replace the heated air at the Earth's midsection. But at approximately 30 degrees north latitude, some of the air has cooled enough to descend to the surface, where it splits again —part to the north, part to the south—and creates the first of

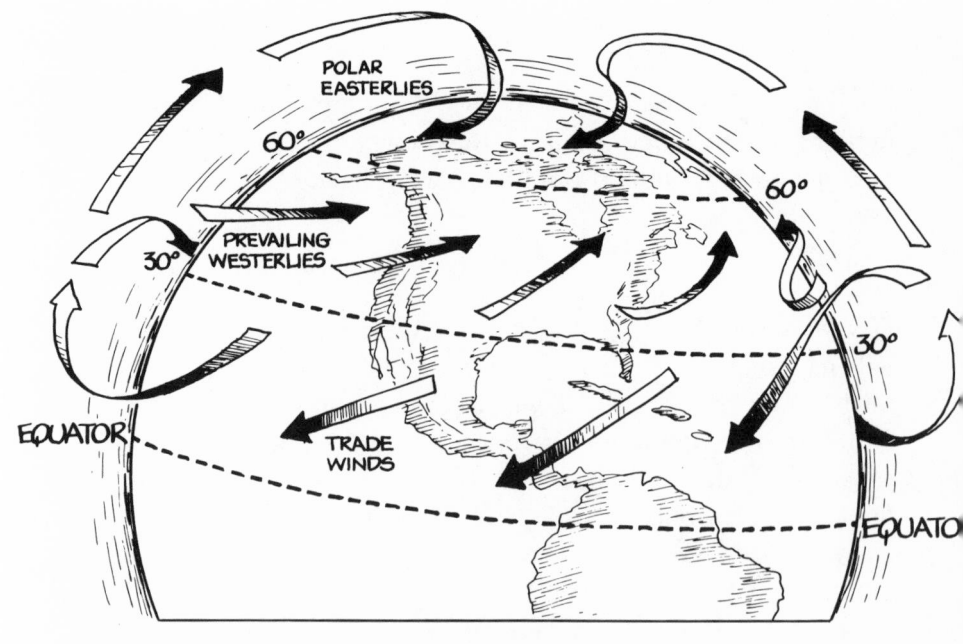

Figure 29

The general circulation of air in the Northern Hemisphere

three cells of circulation. The second cell occurs at between 30 and 60 degrees north, and the third is found in the topmost 30 degrees of latitude.

Based on this theory of heating alone, you would expect to find northerly surface winds from the equator to 30 degrees north, winds out of the south for the next 30 degrees, and prevailing northerlies in the polar cell, but another force comes into play as the result of the rotation of the Earth. Called Coriolis force, in honor of the French scientist who first explained it, this phenomenon causes everything that moves over the Northern Hemisphere surface of our spinning planet to be deflected to the right. Wind is no exception, so in the first cell of circulation the winds that theoretically flow from due north swing to the right and become the northeasterly *trade winds*. In the next 30 degrees of latitude, Coriolis force transforms the southerly winds into the *prevailing westerlies,* and the *polar easterlies* are the primary surface winds in the northernmost regions.

From an atmospheric-pressure standpoint, the general pattern of circulation produces a low-pressure area at the equator, a band of higher pressure around the Earth at about 30 degrees north where the first "sink" occurs, another belt of low pressure at 60 degrees, and at the top of the world a strong high-pressure area in which the coldest air descends to the surface. Of course, these pressure patterns are not at all precise and orderly; they waver back and forth with the seasons and the migrations of air masses. The locations of high- and low-pressure areas and their relative strengths are measured by sampling surface air pressures—higher pressure indicates more air stacked up above the observer, lower pressure means less. When these reports are developed as lines connecting points of equal pressure, a two-dimensional chart of the atmosphere results. From the smallest closed loop outward, the lines (called *isobars—iso* for "the same" and *bar,* shortened from "barometric," for "pressure") indicate the horizontal shape of a "high" or a "low," and the spacing of the isobars (which always represent equal pressure intervals) describes the slope (or topography, if you will) of the pressure system.

There are no numerical scales that make one system a "high" and another a "low," but when weathermen can identify a portion of the atmosphere that is obviously exerting higher pressure than the surrounding air, that area (usually the center of an air mass) is labeled a "high" and the lower-pressure regions around it are called "lows." If they could be seen in 3-D, each high would appear as a mountain of air and each low would turn out to be a depression in the atmosphere, with the air constantly moving from high to low. (A strung-out high is referred to as a *ridge* of high pressure; an elongated low is a *trough* of low pressure.)

Coriolis force exerts itself on the air flowing downhill from a

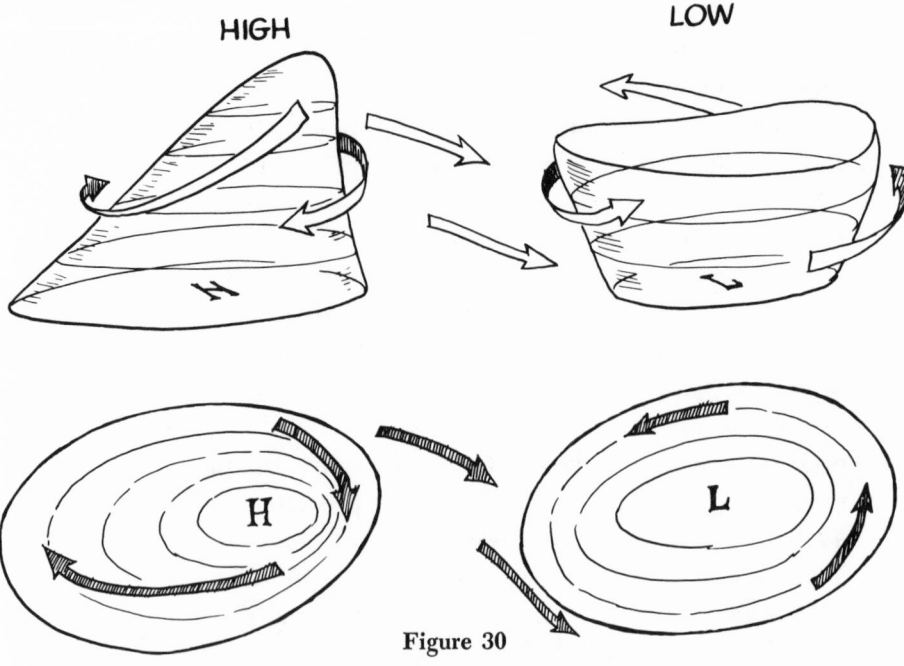

HIGH

LOW

Figure 30

Circulation of air (wind) around high and low pressure systems. In addition to the circular movement, air always moves from areas of high pressure into areas of low pressure

high to a low, with the result that in the Northern Hemisphere winds blow clockwise around highs and counterclockwise around lows; in addition, the winds tend to flow parallel to the isobars. So the pressure-pattern chart is filled with wind information; you can determine the general direction *and* the relative velocity, because the air will obviously flow more rapidly when the air-pressure difference per unit of horizontal distance is greater, as indicated by closely spaced isobars.

The clockwise/counterclockwise principle is bent just a bit in the nap of the Earth, where surface friction tends to slow the wind and change its direction somewhat. This effect is generally noticed from the ground up to about 2000 feet. From there on, relatively free of surface friction, the wind direction increases in bearing (you'd have to turn right to keep the wind in your face) and velocity. As the altitude increases, the general circulation pattern asserts itself even more; it's exceptional to have anything but westerly winds over the United States at 10,000 feet or more above sea level.

In the absence of strong general-circulation winds, certain types of terrain and proximity-to-water situations can develop local winds of significant strength and consistency. Probably the most prevalent of these conditions is the land/sea breeze, which occurs close to large bodies of water. At the heart of the phenomenon is the fact that land areas heat much more rapidly during the day than water does. The heated air rises over the land, and the replacement from over the water creates a *sea breeze*, which can grow into a rather stiff wind during the course of a hot summer day. Unless a weather system overrides it, the process is usually reversed in the evening, since the water retains a great deal of its heat (for the same reasons that make moist air less stable), and the local wind turns into a *land breeze* flowing from shore to sea.

Valley winds represent another local air-movement condition, and also are related to the up-welling of sun-heated air. In this case, air near the ground moves up the slopes and is replaced by cooler air sinking into the center of the valley. At night, especially on a clear night, when the Earth radiates large amounts of heat into the atmosphere and cools the adjacent air considerably, the cooler and heavier air begins to slide back down the sides of the valley and creates a slowly rising fountain of air at the center.

Whenever air moves to a lower level in the atmosphere, some heating due to compression is bound to occur. On occasion, the altitude and temperature change is great enough to set up a *downslope wind* condition, which feeds on itself, the heated air pulling fresh air in behind until the temperature and velocity rise far above normal levels. Sometimes these winds

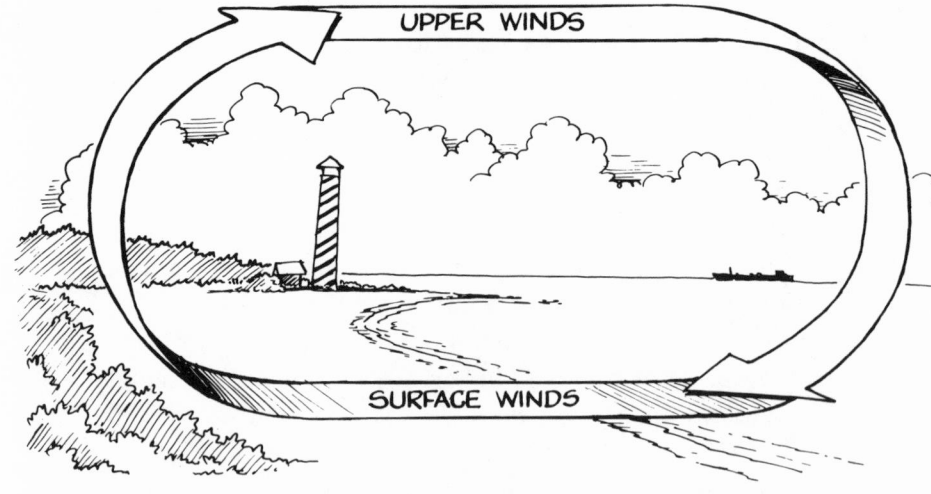

Figure 31

Sea breeze

are mere annoyances, with blowing dust and rough flying, but when the super-warm currents move across snow-packed mountain slopes, the sudden rise in temperature can result in snowslides and flooding. The downslope or mountain wind is predictable in certain parts of the country, and has established itself as a seasonal tradition. On the eastern slopes of the Rockies it's called a Chinook, and the warm wind off the Sierras is known as the Santa Ana.

The super-winds—hurricanes and tornadoes—stand in a class all by themselves. Although a unique set of circumstances is required for the formation and development of a hurricane, the net result is a very deep low-pressure system being filled by warm moist air, which tends to accelerate the vertical and rotational movement of the storm. When the wind velocity in such a weather system reaches a sustained 65 knots, the storm is officially classified as a hurricane. Winds well in excess of 100 knots are not uncommon, and the extremely heavy rain and turbulence must be experienced to be believed.

A tornado is the *coup de grâce* from a granddaddy of a thunderstorm. It results from the swirling of the air ingested by that highly localized energy machine. With awesome low pressures at its center, the rotational velocity of a funnel cloud can exceed 200 knots—strong enough to drive pieces of straw into telephone poles, upset houses and cars, and obviously make short work of an airplane whose pilot happened to get into the path of the storm. Just as every cumulus cloud is a potential thunderstorm, every thunderstorm is a potential tornado producer—another good reason to give all such systems the undisputed right of way in every encounter.

FOG

Fog is nothing more or less than a stratus cloud in contact with or very close to the ground, but the word *fog* in a weather report is usually sufficient to send VFR-limited pilots home to wait for a better day, and even instrument-rated flyers take a second look at a foggy forecast. Fog usually is completely devoid of turbulence and seldom extends very far vertically. The big problem with fog is the restriction to visibility it represents—as often as not, a *total* restriction, and nearly always *very* close to the ground.

The several types of fog share a common denominator. Fog will not form until and unless the air temperature approaches its dew point, precisely the same conditions required to produce any sort of a cloud. A temperature–dew point spread of 3 to 4 degrees Fahrenheit is considered the brink of fog formation; it's almost assured if the air contains particles upon which condensation can occur and there's a light breeze—not more than 10 knots or so—to move the cooling air about. The breeze not only helps create fog, but increases the depth and breadth of its coverage.

Probably the most common type is *radiation* or *ground fog*. The cooling process that creates it begins when the sun sets on a clear night. As heat is radiated back into the atmosphere, the Earth and the adjacent layers of air are cooled. If the cooling continues until the temperature of the air close to the Earth reaches its dew point, fog will form. Radiation fog will usually disappear when the morning sun once again raises the temperature of the air above the dew point.

The development of ground fog is hindered by cloud layers, which reflect some of the radiant heat back to the Earth, much as a giant greenhouse would. The probability of radiation-fog formation is much greater on a clear night, and especially if the starry skies appear following a heavy rainstorm; evaporation from standing water raises the dew point, accelerating fog formation.

Advection fog occurs when moist air and cool surface meet, but in this case the cooling process takes place because of movement: Moist air is carried over relatively cold land or water, condensing as it goes. Advection fog will continue to spread so long as the generating wind blows, and has been known to affect vast areas of the country when a strong on-shore flow propels moisture-laden air far inland over ground still cold from winter. Not as readily burned off by the sun because of the continuing moisture input, advection fog is likely to remain until the wind shifts.

When valleys are clear but the mountaintops and ridges are obscured by stratus-type condensation, the culprit is often *up-slope fog,* produced when moist air is lifted gently to higher elevations. True to the basic principle of fog formation, upslope fog forms whenever the altitude change and the resultant expansion drop the air temperature to its dew point.

Rain falling through cool air—particularly in the vicinity of a sluggish, stagnated weather system with little or no wind—sometimes induces fog formation by the dual effects of evaporation: cooling of the air and addition of moisture to air that may already be close to saturation. *Precipitation fog* generally shows up as very low, dirty gray stratus clouds known to pilots as "scud"—by any other name, just as much a restriction to visibility.

Fog problems exist on occasion when the air gets *very* cold—on the order of 25–30 degrees F. below zero and colder—and when there is no wind blowing. Moisture vapor added to such frigid air may freeze instantly into tiny ice particles light enough to remain suspended in the calm air. The first pilot down the runway may generate a solid trail of *ice fog*, one runway wide and as deep as his wake turbulence. The source of the water vapor is the engine exhaust, spewing out moisture as an inevitable byproduct of combustion. The "instant IFR" thus created may last for some time if the air doesn't move, so the smart cold-weather pilot gets up early to be sure he's the first one off.

FRONTS

If air masses could be colored throughout—blue for cold air and red for warm—the theory of weather fronts could be proved or disproved. As it stands, there is little choice but to accept the meteorologists' theory that a zone of discontinuity exists between unlike air masses, and that they conflict along these boundary zones in a "battlefront" situation.

Frontal theory is well illustrated by the northernmost cell of the earth's circulation pattern, the one that pours cold air down onto the North Pole. All that heavy, dense air has to spread southward (Is there any other direction from the North Pole?), and the leading edge of this bubble of air, probably much like a drop of water of a waxed surface, is known as a *polar front*. When the sun is low in the winter sky and doesn't heat the northern reaches of the Earth very effectively, the polar front encounters little resistance and invades far to the south. Polar outbreak, Blue Norther, cold snap, call it what you will—an

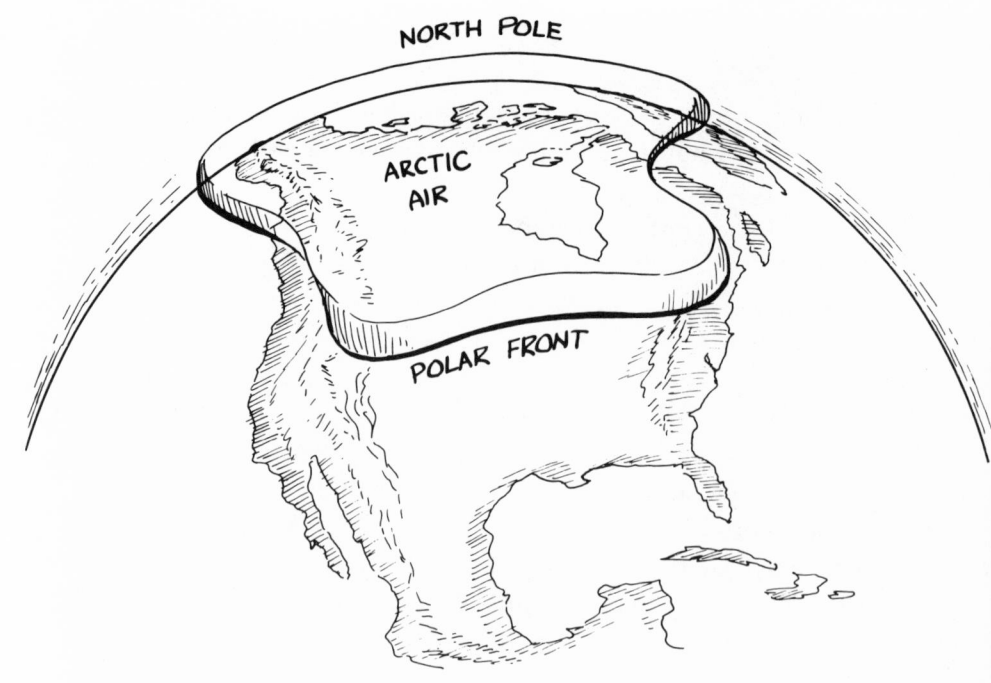

Figure 32

Northern polar front

Arctic air mass under a full head of steam pushes the warm air out of its way clear to Pasadena sometimes. During the summer, when the sun's rays strike the Earth at a high angle, the polar front is beaten back by the pressure of warmer air.

As air masses advance and retreat, the weather produced along the front can vary from disastrous to nothing at all. Changes in temperature, pressure, and humidity alter and affect the air on both sides of this meteorological war zone and generate clouds, fog, precipitation—all the forms of weather known to man. In addition, there are some mechanical considerations involved with the movement of a front, because a bubble of high-pressure air muscles other air masses out of the way, with some of the displaced air inevitably being forced upward—expanding, cooling, forming clouds if there's enough moisture.

When you know something about the air masses involved—relative temperatures, pressures, and moisture content—it's possible to predict with at least a fair amount of confidence the type of weather to be expected where the air masses meet. Conversely, your observation of current weather conditions will often help you identify the frontal situation and come up with a general idea of when the gray skies are gonna clear up.

The classic development of a frontal system starts with a straight boundary line between two unlike masses of air, the colder air on the north side of the line, warmer air to the south. Wind, if any, would be blowing parallel to the front and in opposite directions on either side. When something (maybe local heating, or local circulation) causes a low-pressure area to form along the boundary, a wavelike bend develops along

Figure 33

A typical frontal system, rotating around a low-pressure center as the entire system moves generally eastward

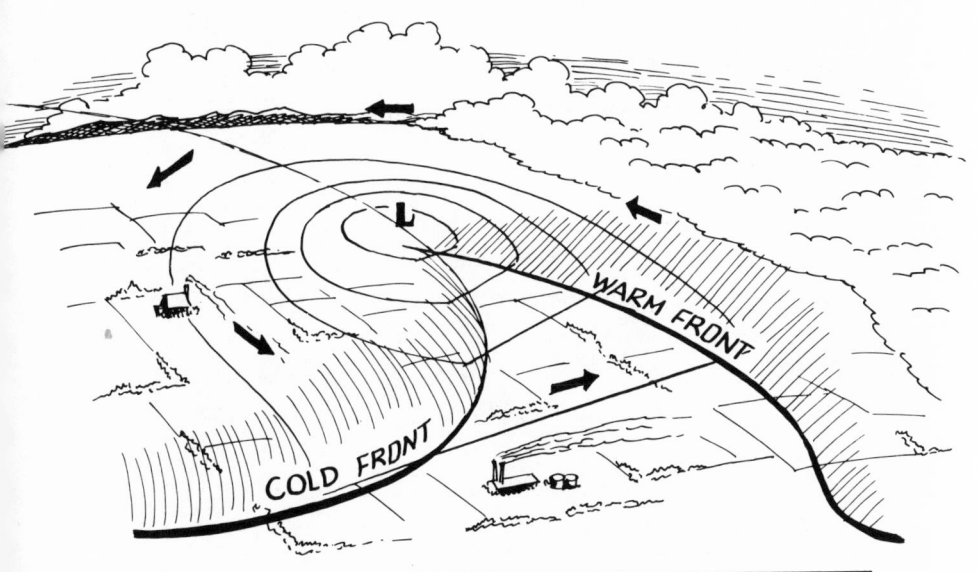

the line, and, with time, two distinct fronts result: a *cold front* at the leading edge of the cooler air mass and a *warm front*, representing the leading edge of the warmer air being pushed out of the way. The entire system moves generally eastward, rotating counterclockwise around the low-pressure area at the center. If at any point along the way the differences in the two air masses are resolved through mixing across the front, the system *washes out* and becomes just another air mass waiting to be moved along by the next front.

Based on prominent characteristics, there are four types of weather fronts—cold, warm, occluded, and stationary:

COLD FRONT: The leading edge of an advancing cold air mass, usually one with high pressure and density but relatively little moisture. Cold fronts are aggressive, and do most of the pushing around in the atmosphere. As the leading edge of a bubble of air, this type of front forces great volumes of displaced air upward: the faster it moves, the more vertical (and violent) weather it generates.

WARM FRONT: The leading edge of a rather flat wedge of retreating cold air. A warm front tends to be more subdued, and moves out of the way of cold fronts.

OCCLUDED FRONT: The result of further development of the wave form in the classic situation, an occluded front represents a mixture of both cold and warm frontal weather. It is produced when the cold front moves *much* more rapidly than the warm front, and either pushes the warm front aloft or rides up and over the sloping surface of the warm front. In either situation, the wave form closes, or *occludes*.

STATIONARY FRONT: When *any* frontal system slows down or stops, the weatherman classifies it as a stationary front. Since the energy of movement is no longer present to generate much weather activity, a stationary front is usually much less violent; the weather conditions will be similar to those associated with a slow-moving warm front.

While the four types of fronts are considerably different in terms of the mechanics of formation and the weather they produce, a group of common characteristics can be traced throughout:

PRESSURE CHANGE: All fronts lie in troughs of low pressure, elongated depressions in the atmosphere. An individual front, or a complete frontal system, always trails outward from a low-pressure center. Whatever it might be in the air mass ahead of any front, the atmospheric pressure drops as the front passes, then increases as the replacing air mass asserts itself. The difference between the pre- and post-frontal pressures (which can be determined by scanning the weather reports from locations on both sides of the front) is a good indicator of the front's speed and of the severity of the weather—if any—associated with it.

TEMPERATURE CHANGE: A drop or rise in air temperature is the most noticeable and reliable indication of frontal passage over a point on the surface, and is regularly used as an indicator by pilots penetrating frontal systems. The temperature change occurs (sometimes dramatically) whether you're on the ground being passed over by a front, or in the air, flying from one air mass to the other. The greater the difference in temperatures, the greater the likelihood of severe weather across the front.

WIND SHIFT: All fronts are accompanied by a change in wind direction, and it is always a clockwise change. (You will need to turn to the right to keep the wind in your face as a front passes.) Sometimes, the wind shift across a weak warm front, is barely noticeable, but a strong, fast-moving cold front will often turn the windsock many degrees while you watch. The amount of wind shift and the severity of weather produced are directly related.

DIRECTION OF MOVEMENT: Because of the prevailing westerly winds in our latitudes, weather systems proceed from west to east, with exceptions that occur from time to time, as in the usual pattern of hurricanes and severe tropical storms, which tend to approach the United States' east coast from the southeast. For normal systems, cold fronts can be expected from a general northwesterly direction, warm fronts from the southwest, and occlusions tend to follow a more direct west-to-east course across the country.

SPEED OF MOVEMENT: The faster a front moves across the ground, the more violent and unpleasant will be the weather it produces. This is especially true of vigorous cold fronts in spring and fall, when the juggernaut from the north is replacing warm, very unstable air. Cold fronts normally move at about 35 knots, with occasional bursts of speed up to 50–60 knots and more; warm fronts amble along at 10–15 knots, and any frontal system that slows to 5 knots or less can be considered stationary. Winter brings a general increase in the speed of fronts, owing to the greater difference in air-mass temperatures. When a front slows down for any reason, the usual result is a flattening of the system, more mixing of the air masses involved, and a general decrease of intensity of the weather being produced.

VERTICAL STRUCTURE: The higher density of colder air dictates that it will always be found underneath the frontal surface,

Figure 34

Typical cold front weather

regardless of the type of front. This generality is important, for the temperature of the colder air will have a lot to say with regard to fog formation and the possibility of freezing rain or heavy snow. For the instrument pilot penetrating a frontal zone, this knowledge is invaluable in determining a course of action if structural icing is encountered.

Their commonalities notwithstanding, each type of weather front has characteristics all its own:

A *cold front* usually produces a relatively narrow band of weather, often a line of towering cumulus clouds where the unstable warm air is bulldozed upward by the leading edge of the cold air mass. Although cloudiness is part and parcel of most frontal situations, it's possible for a "dry front" to develop, with wind shift and temperature-pressure change the only indicators of frontal passage. The precipitation that almost always accompanies a cold front is likely to be localized and showery in nature—large drops, heavy rain, or snow that is over with in a short while. As the front passes, the drop in temperature is often dramatic, and the pressure may fluctuate rapidly. When the aviation weather reports indicate "PRESFR" (pressure falling rapidly) or "PRESRR" (pressure rising rapidly), it's a good bet that a strong, fast-moving cold front is to blame. Southwest winds precede a cold front, generating considerable turbulence at all levels; the wind speed ahead of the front is a good indicator of the severity of the weather following not far behind. Because of the bubble of high-pressure air that generated the cold front, you should expect rapid clearing conditions when the front has passed. If you live and fly in the lee of a large body of open water, beware of the deck of stratus clouds that often forms behind a cold front as the result of moisture evaporating into the cold air. The same principle—surface evaporation—frequently causes early-morning fog when the showers of last night's brisk cold front left water standing on the ground.

Warm-front weather is very nearly opposite in character. A wide belt of passive, gentle weather conditions, it's often so wide that the front itself is hard to find. The cooling of moist air as it climbs slowly up the shallow slope of the front forms clouds with little vertical development, and the widespread, small-droplet precipitation that is the hallmark of the warm front results. (When the lifted air happens to be somewhat unstable, the warm front is capable of producing its own brand of severe thunderstorms, rendered more hazardous because they're buried in the heavy clouds near the base of the front.) Rather than being the abrupt change from hot to cold that is often the case with a cold front, the temperature difference across a warm front is more likely to be characterized as a

WARM AIR

COLD AIR

Figure 35

Typical warm front weather

change from cool to warm—and it doesn't take place rapidly. Winds ahead of a warm front are much gentler, generally from the south or southeast. Given the wide band of weather common to warm fronts, the clearing of the air after frontal passage is very slow and gradual. The relatively flat profile of a warm front also accounts for the small change in atmospheric pressure on either side of the front.

Occluded fronts produce a mixture of the weather phenomena that characterize the other types. There may be locally heavy precipitation inherited from the cumulus development of the cold front, and far-flung stratus and low visibility conditions as a result of the warm front's influence. An occlusion generates its worst weather during the initial stages; as time goes by, energy is dissipated, and the eventual mixing of the two air masses results in a dissolution of the frontal zone.

The *stationary front* is the quick-change artist of the atmosphere. With weather much like that of a warm front but less severe (lacking the energies of movement), the band of predominantly stratus clouds and drizzle spreads out over a wide area, perhaps moving a bit north during the day in response to

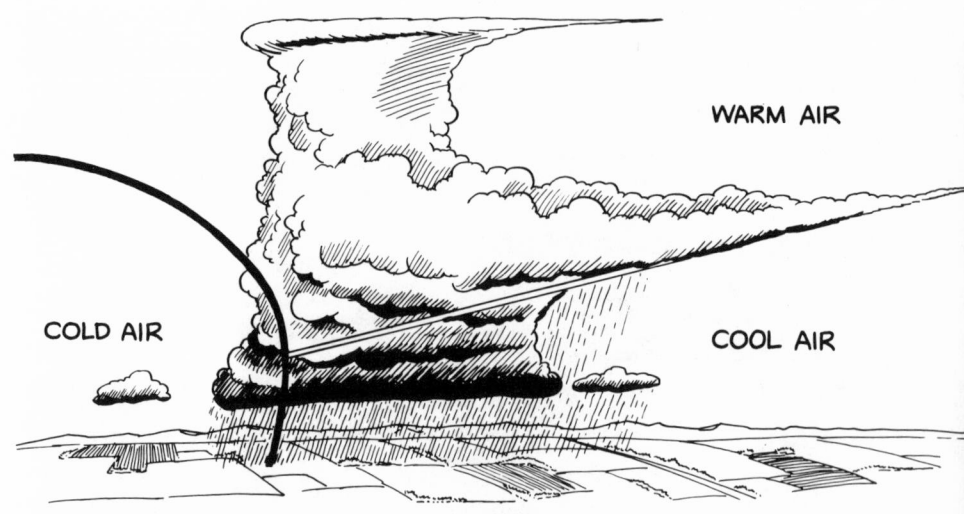

WARM AIR

COLD AIR

COOL AIR

Figure 36

Typical occluded front weather

the pressure of sun-heated air, then moving southward when the sun sets and cooler air exerts the greater force. A persistent stationary front can vacillate back and forth for days on end, with low ceilings and visibilities over many thousands of square miles. The stationary front is to be expected frequently in the spring and fall, when the battle of cold northern air with warm tropical air is a meteorological standoff. This equalization of atmospheric pressures tends to take place in mid-continent.

AIR-MASS WEATHER

After a front passes (or after you've flown through one), weather conditions depend mostly on the moisture content of the air mass and the heating or cooling that may occur. A moist air mass over land of a lower temperature will be cooled from below, and you should expect low ceilings and visibilities, the degree of which will be a function of the dew point–temperature relationship. On the other hand, a moist but clear post-frontal air mass that is heated by the surface will likely

produce localized cumulus clouds—"fair-weather cumulus"—
and scattered thunderstorms when gross instability is present.
When an air mass is very dry, look forward to fine, bright
weather on the west side of the front.

In those areas where the atmosphere is not fouled by smoke-
stack emissions and the inevitable byproducts of energy production, the crisp, clear air behind a cold front will exist for
days—often until the next weather system moves through—but
those conditions seldom exist around our large urban centers.
Within a day or two after cold frontal passage, man-made pollutants begin to cloud the clear skies and provide the nuclei for
condensation. This is industrial modification of an air mass,
and frequently creates low visibility in the form of a haze or
smog layer that can extend upward of 10,000 feet above the
surface. So far as pilots are concerned, the source of the restriction is immaterial—low visibility is low visibility is low visibility.

When an air mass comes to a standstill, it begins to take on
the characteristics of temperature and humidity that exist on
the surface. Significant modification can occur when an air
mass remains over a particular land or water area for an extended period of time: Moisture content may increase; temperature may rise or fall and provide a completely different set of
results when the air mass is finally pushed out of the way by
the next system.

In certain situations, the normal lapse rate reverses and
temperature actually rises with altitude. Known as an *inversion,* this phenomenon is usually the result of a flow of warmer
air aloft or the contact cooling of the layer of air next to the
surface. For most flight operations, an inversion represents
nothing more than an interesting quirk in the atmosphere, but
warmer air aloft often provides a rather effective "lid" that
inhibits the lifting and subsequent redistribution of pollutants
throughout the atmosphere. The result is usually a concentra-

tion of debris, which lowers visibility and sometimes creates a health hazard for the entire population in that area. Those are the days when it's a genuine pleasure to fly above the dirty brown layer of smog and fill your lungs with clean air, but alas, we must all too soon return to earth, inhale the gunk like everyone else, and wait for another weather system to clean up the atmosphere.

HAZARDOUS WEATHER

Thunderstorms, the angriest manifestation of atmospheric fury, are capable of containing elements of all the other forms of hazardous weather. Long the most-feared, most-respected weather phenomenon, thunderstorms present to the instrument pilot all the external hazards plus those found *inside* the storm; for VFR fliers, certain hazards exist well outside the storm.

Aviation educators have for so long taught and warned of the horrible things that can happen in and near thunderstorms (use of descriptors such as "awesome" and "the intimidator" is not infrequent) that the question must be asked, "Are thunderstorms *really* as bad as they say?" The answer is an unequivocal YES—and then some. You may flirt successfully with thunderstorms for most of a flying career; then the big one, the one that's bigger than you or your equipment, comes along. Thunderstorms are awesome, unpredictable, intimidating, and should be given a wide berth. It takes only one.

Thunderstorms can be expected to occur whenever the air is sufficiently unstable and some sort of adequate lifting action takes place. The steep slope of a rapidly moving cold front is a prime producer of the lifting action that triggers heavy thunderstorms in advance of the front; rising terrain is another culprit. All year long in the South, and during the summer months in most of the rest of the United States, uneven heating of the Earth by the sun creates convective currents, which lift unstable air thermally and set off thunderstorms.

The growth process of a storm is predictable and usually visible, and occurs in three stages: cumulus (building), mature, and dissipating. In the cumulus stage, unstable air has been lifted and a cloud with vertical development forms; all air movement is upward, precipitation has not yet started, and the net result of flying through such a cloud is a healthy boost upward. Sooner or later, the droplets of water inside the cloud become so large and heavy that they can no longer be borne up by the vertical currents of air, and rain begins to fall, dragging air along with it.

The line between the building stage and full maturity is a very thin one, but when the storm is fully developed there are airplane-bending combinations of adjacent up- and downdrafts inside the cloud, heavy precipitation under the storm, and great heavings of the atmosphere that may disturb the air for miles around. The mature stage is universally recognized as the most dangerous; all the hazards will be at their worst when this condition is attained.

The dissipating stage of a thunderstorm's life cycle begins when the energy inputs slow down. Perhaps the cold front slows, until the lifting action is insufficient to keep things going, or the sun sets and removes the heat source that started the whole process, or the storm itself moves with prevailing winds and drifts away from the hot spot.

These three stages—building, mature, dissipating—are easy to recognize when you can visually isolate one storm cell and follow it throughout its life, but a fully developed storm system contains numerous cells in various stages, and therein lies one of the inherent dangers—you can never be completely certain that the building cell in front of you isn't hiding another in the full bloom of maturity.

Hazards outside the storm—those of greatest concern to the

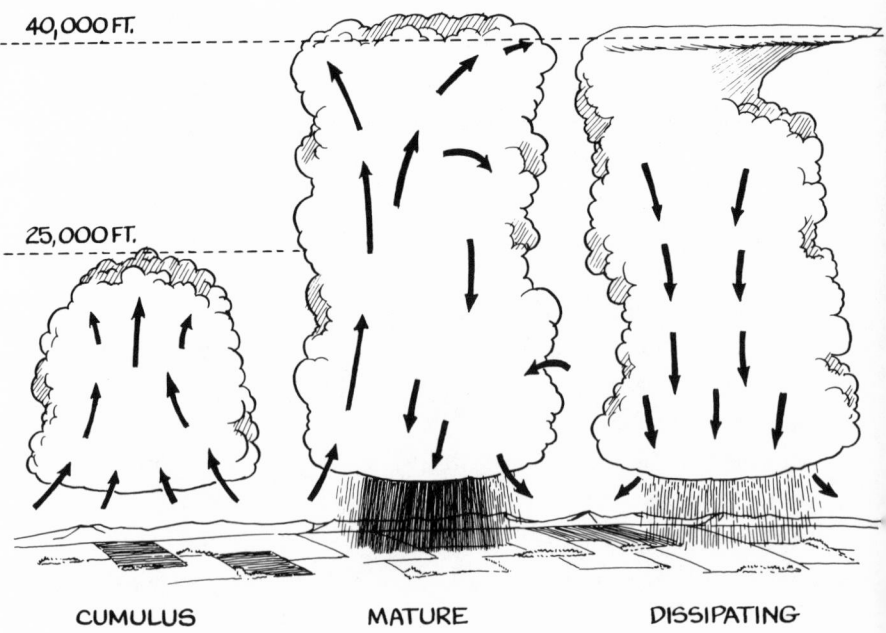

40,000 FT.

25,000 FT.

CUMULUS   MATURE   DISSIPATING

Figure 37

Stages in the life cycle of a thunderstorm

fair-weather pilot—include high speed, gusty surface winds that can change direction rapidly, heavy turbulence in the clear air for several miles around the storm, and heavy rain, which lowers visibility and frequently masks other storm cells. Hail is produced by most thunderstorms and, if thrown out the top of the cloud, can fall in the clear air downwind of the storm. Flying into a flock of ice cubes is a surprise at best, a disaster at worst. A tall, broad thunderstorm (height is the best indicator of intensity) shuts out some of the light of day, hastens the onset of darkness, and causes problems for the pilot who is determined to get home before the lights go out. On top of all these unpleasant features of thunderstorms, the possibility of invisible tornadic tubes—now known to circulate inside major storms, and suspected of existing in the clear air *between* adjacent thunderheads—makes it easy to advise pilots about flying near thunderstorms. *Don't.*

Because the conditions required for thunderstorm formation are so simplistic—unstable air plus lifting action—it's not difficult to predict when they are likely to occur. Any time the local atmosphere is warm and moist, the potential has been halfway satisfied; add some kind of lift, and cumulus activity will result. *Every* approaching cold front should be immediately suspect, with the fast-moving ones deserving even more attention—they often exert their influence hundreds of miles ahead, in the form of an instability line (much like the bow wave of a boat), which generates *squall line storms,* the most violent of them all.

Freezing rain happens several times each winter in the northern parts of the United States. Most likely, a warm front with a renewed lease on life pushes northward; the overrunning warm air produces rain, which falls through the wedge of below-freezing winter air near the surface. By the time it reaches the ground, the rain has been super-cooled (it is still in liquid form, but at a temperature below freezing) and will freeze instantly to whatever it touches—trees, sidewalks, cars, and airplanes. It's often possible to fly safely and legally in the ceiling and visibility conditions that accompany freezing rain or drizzle, but it's seldom very smart to do so. Serious hazards exist in the form of reduced visibility when the windshield ices over, and the additional weight the ice represents; a light airplane can suddenly become much heavier when exposed to freezing rain. A heavier airplane will stall at a higher airspeed, requires more power to maintain flight (don't forget that the propeller will be coated with ice, too), and may develop unusual handling characteristics because the ice coating has distorted the normal airflow over wings and control surfaces.

When conditions are ripe for freezing rain, the best place for the aviator is at home, with his feet in front of the fire—even the highways are treacherous in an ice storm. Freezing rain is not so difficult to predict that you shouldn't have ample warning of its approach. Don't plan to take off until the rain has stopped, and if you should fly into a shower of "liquid ice," land as soon as possible, doing everything very carefully and with a bit of additional airspeed to compensate for the effects of ice on the airfoils.

Structural icing is classified according to its physical appearance: *rime ice* is rough, cloudy, light in weight; *clear ice* is smooth, transparent, and very heavy. One or a mixture of both types of icing can form on the outside surfaces of an airplane operating in visible moisture (clouds) when the air temperature is at or below freezing. Since one of the requirements is flight inside clouds (except for the freezing-rain situation), structural icing is a concern limited to instrument pilots, but there's one type that can shoot down the VFR folks as well: *frost,* which is nothing more than rime ice deposited on the upper surfaces of an airplane when it sits outside in moist air and cold temperatures. The rough surface produced may inhibit lift production to the point at which takeoff is impossible, let alone continued flight. The rule is simple: Never attempt to fly an airplane until all the frost has been removed, preferably by placing the machine in a heated hangar.

High winds—at a sustained velocity or in the form of gusts —present a formidable hazard to aircraft, and the lighter the aircraft, the more dangerous the wind. In addition to the obvious problems of aircraft control when the wind is blowing faster than the stall speed (the airplane won't stop flying), the vertical differential in wind speeds creates its own special characteristics. If air moved rapidly without stumbling there would be no problem, but whenever the wind blows faster than 15 knots or so, it is slowed significantly by contact with the ground; expect a sudden increase in wind velocity right after takeoff, and an equally sudden decrease as you approach the runway for landing. Sailplane pilots compensate by approaching at a higher airspeed; power pilots have additional thrust available to make up for the sudden loss of performance.

Gusty winds are uncomfortable and distracting, but the big problem is the rapid change in the angle of attack when a vertical gust is encountered. Given the normal relative wind, parallel and opposite to the flight path of the airplane, a gust of wind that is directed even slightly upward will subject the wings to an immediate, though short-lived, high angle of attack. Some lift is lost, and you'll feel a sudden drop. Carried to extreme, the momentary high angle could cause the wing to stall—if this should happen close to the ground, a very hard landing (or worse) might result. A prudent pilot flies at a slightly higher airspeed when the final approach path is populated by gusty winds.

WEATHER REPORTS

The best ones are yours—even if they amount to nothing more than wetting a finger to test the wind and a quick look at the sky to see what's going on meteorologically—but personal weather reports aren't good enough for anything but the immediate local area. For a flight beyond the reach of your eyes, the National Weather Service provides hundreds of comprehensive reports of the weather observations at all airline terminals and many smaller airports. Short of actually being there, this is the best pilot-weather information available.

Regardless of the source, any weather report is necessarily *history*—it's a report of weather that has already taken place. Trained observers record the measurements of those meteorological qualities of interest to the aviation community, then disseminate the information via teletype to all Flight Service

Stations and to anyone who cares to pay for the service; it's available to all comers on a lease basis. Most nonprofessional flyers never see real teletype weather reports after they go to school for written pilot examinations, but it's important to maintain at least a general idea of the content so that you'll understand what a weather briefer is telling you over the phone, and so that you can ask intelligent questions about the weather situation.

HOURLY AVIATION REPORTS

A standard format is used, the sequence beginning with the three-letter identifier for each station at which an observation was recorded. (It pays to learn which airports issue hourly reports in the area you travel most frequently; when you're concerned about the weather at a non-reporting airport, you can get the information for nearby terminals and interpolate the conditions.) The cloud report is next, all heights presented in hundreds of feet above the ground, and the extent to which the clouds cover the sky is indicated by either CLR (clear—no clouds at all, or less than 10 percent coverage), SCT (scattered —up to 50 percent of the sky is covered), BKN (broken—more than 50 percent but not more than 90 percent of the sky is covered by clouds) or OVC (overcast—more than 90 percent coverage). An instrument clearance is required to penetrate a broken or overcast layer, both of which are officially designated *ceilings*. When smoke, dust, snow, or rain prevent observation of cloud cover, the reporter will indicate the obscuration with an "X." On occasion, cloud cover may be reported as "thin," which indicates a transparent layer.

Surface visibility is reported next: the greatest distance (in statute miles) the observer was able to see over at least half the horizon. This number is often in disagreement with the visibility in other directions from the observation point, but it will always reflect the worst conditions observed. (For the more critical requirements of instrument operations, visibilities of less than one mile are frequently determined by the installation of a specialized photo-electric sensor mounted close to the runway to provide tower controllers with a digital readout in hundreds of feet.) Significant restrictions in the form of precipitation or other atmospheric conditions are reported immediately after the visibility value: Rain (R), snow (S), thunderstorm (T), dust (D), fog (F), haze (H), smoke (K) are examples.

The next item is sea-level pressure in *millibars*, the weather observer's metric measurement of the weight of air above him. This coded report (only the last three digits are shown; the third one represents tenths of a millibar) is of little interest to pilots, but by comparing several recent hourly reports you can determine how rapidly the pressure has gone up or down. The same information is provided at the end of the report in *inches of mercury*—also presented in a three-digit code, the second and third numbers representing hundredths of an inch. Both air-pressure reports can be used for altimeter settings.

Temperature and dew point are reported in degrees Fahrenheit. A look at the past few hourly reports will confirm the rate at which these two measurements are closing or spreading, a good index of the potential onset or dispersal of visible moisture in the air.

Winds measured at the surface are reported in a code that omits the third digit in the wind direction: "2412" means the wind was blowing from 240 degrees (always measured with reference to true north) at a velocity of 12 knots. When the wind is gusty, the letter "G" is added, followed by the highest velocity noted by the observer: "2412G26" is the same prevailing wind as before, with the addition of a 26-knot gust.

In the remarks section of the hourly report, the aviation world discovers the more intimate details of the weather situation; the reporter fills in the basic information with such things as variable visibilities and cloud conditions, thunderstorm location and movement, pilot reports of weather phenomena observed from the air, and just about anything that will enhance the interpretation of the weather situation, plus Notices to Airmen (NOTAMs), a system of informing pilots of runway closings, VORs out of service, changes in airport lighting systems, and a thousand other items of an operational nature that could affect the conduct of a flight.

A typical hourly aviation weather report would look like this: APR 7SCT M18BKN E200OVC 6KH 129/60/59/2504/991. At the APR Airport, there were scattered clouds at 700 feet (the "00" is always omitted), a measured ceiling of broken clouds at 1800 feet, and an overcast layer estimated at 20,000 feet. Visibility was six miles in smoke and haze, the sea-level pressure was 1012.9 millibars, temperature and dew point were, respectively, 60 and 59 degrees Fahrenheit, the wind was blowing from 250 degrees at 4 knots, and the altimeter setting was 29.91 inches of mercury.

OTHER WEATHER REPORTS

Flight Service Stations and National Weather Service offices assemble a wealth of weather information in the process of providing the boiled-down version to pilots. Most of the reports are of little use by themselves, but there are two that can be of inestimable help in flight planning when the weather is less than ideal: the *weather-depiction chart* and the *radar-summary chart*. Even if you don't have the opportunity to see these charts, the briefer can relay a great deal of decision-making information over the telephone or radio.

The weather-depiction chart is a graphic summary of the hourly aviation reports, and highlights those areas where instrument conditions existed at report time. By examining the previous chart or two, the briefer can determine which way the IFR weather is moving and so help you plan your flight to avoid it. For instrument pilots, the weather-depiction chart is perhaps most useful for indicating which way to turn to fly *out* of instrument conditions should all the radios fail in flight. It's smart to have an "out" for every possible meteorological trap.

The radar-summary chart is similar to the depiction chart, with areas of observed radar echoes indicated. Not only does this chart show where precipitation is likely to exist (radar "sees" rain and snow), it also helps answer your questions regarding the vertical extent of such areas—the heights of radar echoes are printed on the chart. In its most useful capacity, the radar summary lets pilots know where the thunderstorms were and, by virtue of the frequent updates when there's a lot of radar-detectable weather around, can help keep you posted on the movement of the violent weather areas.

WEATHER FORECASTS

The National Weather Service provides information about future aviation weather with *national* forecasts covering the entire country, *area* forecasts covering several states, and *terminal* forecasts for specific airports.

The national forecast chart (in graphic form) deals with the expected positions of major weather systems—highs, lows, fronts—and the cloud cover and precipitation that will accompany them. This "big picture" of the national weather scene is presented four times each day, each forecast looking ahead for 12- and 24-hour periods.

Somewhat more detailed, because it covers a much smaller chunk of real estate, the area forecast is produced in verbal form and transmitted via teletypewriter. The standard contents deal with the position and expected movement of weather systems, cloud bases and tops, visibility over the area, turbulence, and icing conditions. Valid for 18 hours, area forecasts are published twice each day.

Terminal forecasts are directed to the weather conditions you should expect in the immediate vicinity of an airport. Issued three times each day, the "terminals" are valid for a period of 24 hours. Within the forecast period, the weathermen will indicate the times at which they expect significant changes in cloud height and coverage, visibility, wind conditions, and precipitation.

Covering a more specific segment of the aviation weather picture, the *winds-aloft forecast* is a completely computer-generated teletype printout of wind direction and velocity expected above 100 major U.S. terminals. Winds (and temperatures) are forecast for nine levels from 3000 to 39,000 feet above sea level, and provide good basic information for flight planning.

WEATHER ADVISORIES

When hazardous weather conditions (tornadoes, squall lines, severe icing, turbulence, and the like) develop anywhere in the country, an appropriate advisory will be issued. Intended primarily for those pilots who are already airborne, advisories are known as SIGMETs when the hazards apply to all aeronautical interests, AIRMETs when only light aircraft will likely be affected. Both types are broadcast on the navigational frequencies for the area of concern, and are also disseminated via teletype.

SOURCES OF WEATHER INFORMATION

The Flight Service Station is the fountain of knowledge for most pilots in the matter of weather information. A personal visit, a telephone call, or an airborne radio contact provides immediate access to the world's most extensive weather-information system, and to a specialist trained to help you interpret whatever reports and forecasts might apply to the operation you propose. In those big-city areas where the Flight Service Station gets far more telephone requests for weather briefings than it can possibly handle, special numbers are listed that provide a continuous, updated recording of the briefings for popular flight routes. When airborne, you are seldom beyond the range of one of the low-frequency radio beacons or VOR stations designated to carry a similar recorded broadcast of the weather conditions existing and a forecast for that general area.

In addition to Uncle Sam's official information, the daily newspaper and the television set offer weather information that is very general in nature and coverage, but can at least give you an insight into the kind of weather you should expect today or tomorrow. Newspapers publish a facsimile of the national forecast that shows fronts, precipitation, the positions of highs and lows—enough to give a student of aviation weather a good idea of the conditions coming up.

TV weather shows at the local level vary from excellent to hardly worth watching, but one national program has consistently presented a weather segment so good that it is watched by many professional pilots: the *Today* show, whose daily forecasts of general weather conditions, temperatures, and clear warnings of hazardous weather areas are to be commended.

LOCAL PHENOMENA

All of the weather situations discussed in this chapter are the classics—the expectations of the weather prophets when things happen according to the book. But, just as likely as not, the modification of an air mass or the route of a major system will be altered remarkably by rising or falling terrain, a large body of water close by, a heavy concentration of industrial activity, or any of a countless number of combinations.

Since you will be flying locally most of the time, it's a good idea to "keep book" on the weather as it occurs, noting the conditions that tend to set up bothersome conditions. The next time the same set of precursors shows up, you'll be one step ahead of the game. And when you find yourself in some far-off part of the country, unfamiliar with the ins and outs of somebody else's local weather, don't expect things to happen the way they do at home, but get in touch with the weatherman who knows the area—in short, "ask the man who's been there."

Aerial Navigation

9

It wasn't always so easy, but today, if you can get on the right interstate highway, head in the proper direction, and keep your eyes on the odometer and the exit signs, you can navigate an automobile from one city to the next with utter simplicity. There's no concern that the wind might blow your car off the road, and with the national speed limit in effect, the trip that required two hours last month will most likely take the same amount of time tomorrow. An automobile's course is determined by the roadway itself, and the distance traveled is right there in black-and-white numbers. Together they provide the highway navigator with absolute information about his position and for his estimates of when he'll be where.

Not so with an airplane, because it is moving through its medium rather than on it, and the medium itself is usually moving as well. When the air moves laterally in respect to the direction you want to fly, a correction must be made or the airplane will certainly drift off course; fore-and-aft movements of the air (headwind or tailwind) will affect the time en route even though the airspeed remains unchanged—sometimes good, sometimes not so good. Add the dimension of speed—even the slowest airplanes move considerably faster than the customary modes of surface travel—and the problems of aerial navigation take on a unique significance.

Pilots have borrowed heavily from marine-navigation methods, particularly "dead reckoning"—if you know the point and the time of departure, you can make an allowance for wind effect and predict where the airplane should be at any given time. Dead reckoning was the backbone of early aerial navigation efforts, and was later combined with celestial observations that made it possible to establish relatively accurate positions en route. While celestial navigation has fallen into almost complete disuse since the advent of radio aids, dead reckoning remains at the core of all preflight problem-solving and inflight corrections.

General aviation navigation methods have boiled down to two distinct situations: Most VFR pilots rely on pilotage, flying from one visible landmark to the next, while instrument pilots use electronic aids as their primary source of position information. As navigational radio signals proliferate in our airspace system and more pilots become aware of their advantages, there is a definite trend toward a combination of pilotage and electronic methods in VFR operations.

ELEMENTS OF AERIAL NAVIGATION

The directional alignment of the proposed course, the number of miles to be flown, and the speed of the airplane through the air are the positive parts of any navigational problem; the wind direction and velocity represent the most changeable quantities, and will always affect the time required for a flight. So *course, distance, airspeed, wind,* and *time* are the basic tools in the aerial navigator's kit.

A *course* is a line drawn between two points on a chart, and is identified by the number of degrees it is displaced from north—a course toward due east is 90 degrees, south 180, west 270, and so on around the 360 points of the compass. The angle is easily measured using a protractor (an integral part of a pilot's navigational plotter), which is placed over one of the vertical lines on the chart. These lines, which pass through both poles, are the scale of east-west distances—longitude—beginning at Greenwich, England. But their importance for Northern Hemisphere navigators lies in the fact that they all converge at the North Pole, and provide the angular measurement of a course with reference to true north.

Unfortunately, the compass in your airplane "looks at" a north that is located some distance from the map-makers' north: The *magnetic* north pole is on Prince of Wales Island, 900 miles south of the geographic pole. At certain locations, the true course you measure on the chart and the angular relationship to magnetic north will be the same, but most of the time you must be concerned about the difference between these two readings, or the magnetic *variation*. Traversing the United States from Boston to Seattle, you would experience a range of variations from 15 degrees westerly (in the Boston area the magnetic pole is about 15 degrees to the west of true north) to 20 degrees easterly in the Pacific Northwest—and in between, the area in which there is no difference, or zero variation, goes from roughly the upper Mississippi valley to Florida.

All aeronautical charts show variation with lightly printed, appropriately labeled lines, so that after measuring a course angle with respect to true north, you can apply the correction to come up with a magnetic course—a number that relates much more easily to the readings from the magnetic compass, your primary source of directional information in flight. Converting true courses to magnetic courses is easy: Whenever an easterly variation is involved, it is subtracted; a westerly variation is always added to the true-course number. A pilots' memory-jogger has grown up around this relationship: "East is least, west is best." So a course line that measures 050 degrees on the chart in an area where the variation is 15 degrees westerly results in a true course of 050 and a magnetic course of 065—the same line is drawn between the same two points, but its direction is expressed in two different languages, true and magnetic.

Distance also is bilingual—nautical and statute. Although there is a strong move afoot to convert all aeronautical readings to nautical miles (the airlines and the military have been working that way for years), the general aviation fleet and its pilots will no doubt hang on to the more familiar statute measurements for a long time to come. The choice is yours, but be aware that you must settle on one or the other. Most charts

Figure 38

A course line plotted on an aeronautical chart is based on true north, and must be corrected for variation. Each *vertical* line is marked in increments of 1 nautical mile

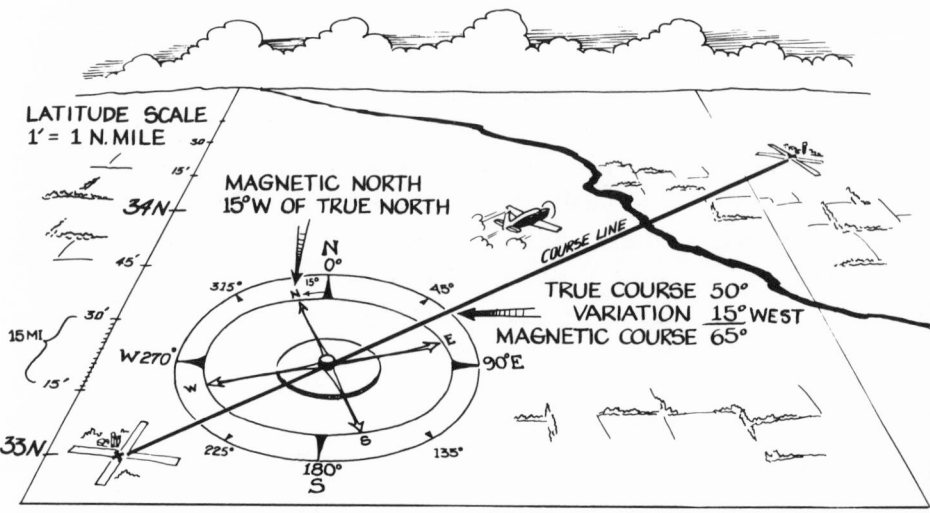

have scales for both, and in some cases for kilometers as well.

The convenience of using nautical miles is heightened by the very definition of a nautical mile: 1 minute (one 60th of a degree) of latitude on the north-south scale of distance around the Earth. You'll find 1-nautical-mile marks on all the vertical lines printed on aeronautical charts; you can't get any more accurate than that, but never measure distances along a *horizontal* line—the scale changes as you proceed toward the pole, and you'll find the marks getting closer and closer together. Use vertical lines when you're looking for nautical miles, and you can find out how far you have to travel by measuring the length of the course line you've drawn (use dividers, a pencil, the edge of a card, anything that's handy) and comparing it to the marks on one of the vertical lines on the chart. (Speaking of anything handy, it turns out that the Sheboygan Airport is exactly 4 finger-widths, measured at the second knuckle, from the airport at Manitowoc, Wisconsin—just about 24 nautical miles, and that's close enough for low-speed navigation.)

Airspeed is one of the most stable quantities with which the aerial navigator works. Once a cruise power setting is determined, there will be precious little change in the number at the end of the airspeed pointer. But you must be aware that as the altitude and/or the airspeed increase, the difference between what you see on the instrument (indicated airspeed) and the airplane's actual velocity (true airspeed) gets larger and larger. The rule of thumb that adds 2 percent per 1000 feet of altitude to the indicated airspeed provides a creditable correction for most airplanes; when more accurate numbers are needed, a navigational computer supplies the answer.

Wind and its effects on the flight of an airplane are what make aerial navigation an imprecise art. The world's finest computers, working with inputs from knowledgeable meteorologists, produce forecasts of wind direction and velocity that are at best approximations for huge volumes of airspace. The pilot who figures out the direction and speed of the wind while he's in flight can only speculate on what those qualities will be a few minutes later, at the same or at some other altitude. Nevertheless, wind forecasts provide a platform from which to estimate the solution of the navigational problem.

Take comfort in the standardization of wind language. All directions, reported and forecast, relate to that point on the compass from which the wind is blowing, and all wind speeds are given in knots—nautical miles per hour. When a tower controller provides wind information, he's speaking of direction in terms of magnetic north, since the runways are designated that way (ATIS winds are considered tower winds). All *forecast* wind directions relate to true north; magnetic variation is not measured above the surface.

The effect of the wind on the flight of an airplane can be broken into headwind/tailwind components and crosswinds, which, if uncorrected, will result in a drift to the left or right of the intended course. If the air through which you're flying happens to be moving in the same direction, you'll have a tailwind component, and the resultant speed across the ground will be the sum of your true airspeed and the velocity of the wind that's pushing you. When you are flying into the wind, the groundspeed—the actual number of miles covered across the ground each hour—will always be the true airspeed minus the speed of the wind.

Groundspeed is the key to successful air navigation; when you know the rate at which your airplane is moving across the ground, you can make an accurate estimate of the *time* between here and the next checkpoint. That's important if for nothing else than to help you decide whether your fuel supply is sufficient for the trip at hand. At a given power setting, the engine consumes fuel at a rate that will empty the tanks in a predictable span of time; a tailwind will boost the miles you can travel during that time, while a headwind may dramatically decrease the distance you can fly before eating into a safe reserve in the tanks. In general, the presence of headwinds should be treated as a warning signal to begin planning alternative courses of action. When the winds are pushing you along the way, be happy, but treat the tailwind benefit as a safety cushion—it's like money in the bank.

When the wind is blowing from either side, an airplane behaves just like a free balloon—that is, it drifts with the wind— unless the pilot takes corrective action. If you could hold your airplane motionless over some spot on the ground and a 20-knot right crosswind sprang up, you'd find yourself exactly 20 miles to the left of that spot an hour later. Even if you resume the normal forward speed of your aircraft under the effect of that 20-knot crosswind, you'd be 20 miles off course for every hour of flight—farther along the course, to be sure, but the drifting effect of a crosswind is inexorable.

Since an airplane is free of contact with the surface, the solution to the crosswind problem is found in turning the airplane so that some of its forward velocity is used to overcome the drifting effect of the wind. The amount of turn is known as the *wind-correction angle*. You can figure it out with simple trigonometry or with a navigational computer. The net result is always a change in groundspeed. If the crosswind is just that— a direct crosswind—or has even a small component of headwind, the groundspeed will be lower; a crosswind/tailwind combination in which the tailwind component is greater than the price paid for drift correction will increase your groundspeed.

After a while, you'll find yourself estimating times and wind corrections, and your estimates will improve as your experience provides a growing foundation of data. You'll also be less upset with the occasional errors that are bound to show up because it *is* an imprecise art. The smart pilot keeps track of his progress through the air, sees the changes developing, and takes appropriate action to keep the little mistakes from growing into big ones.

AERONAUTICAL CHARTS

For visual operations, you can use any kind of chart (map) at hand—even a road map from the neighborhood gas station is good enough for simple pilotage—but the charts built specifically for aviators are much better. They have more detail where it counts, and a wealth of items intended just for pilots: color coding and symbols to indicate the various types of airspace, communications frequencies, airport layouts and other information, and the locations and frequencies of all radio navigational aids, to name a few.

The most popular aeronautical chart is the *sectional chart*. Thirty-seven of them cover the continental United States, each chart showing more than 100,000 square miles of the Earth's surface on a scale of 1: 500,000. In addition to cultural features (cities, towns, railroads, highways, even drive-in theaters and racetracks), the sectional chart depicts lakes, streams, and elevation of the terrain. Sectionals cost at least $1.85 (like everything else, the price of knowing where you are keeps going up), and are available from airport operators or directly from the government.

Another visual-type chart at twice the scale—1: 1,000,000— is the *World Aeronautical Chart*, or WAC. Intended for faster airplanes or longer trips, this series requires only 11 charts for the entire country and provides slightly less detail over a much larger area. WACs are available from the same sources as sectionals, and also cost $1.85 apiece.

Sectionals and WACs are the day-to-day working charts for most pilots. For those who need larger- or smaller-scaled presentations, the government prints *local aeronautical charts* for the big cities and a *nationwide planning chart* that helps you get set for a very long trip or, when framed, makes a great conversation piece on the wall of your office.

The navigational charts intended for use in the instrument system are vastly different. The most notable change is the absence of color, since terrain features and city boundaries are not shown. Airport symbols are there, and major shorelines, but the purpose of the IFR charts is to depict electronic aids to navigation. It's assumed that the users will be buried deep inside the clouds, with no need for visual references.

The instrument world is actually two layers of airspace—the lower deck, up to 18,000 feet and the upper portion, from there to 45,000 feet—with a series of charts for each. The lower levels are covered by 28 *Low-Altitude Enroute Charts*, and above-18,000-foot aviators use four *High-Altitude Enroute Charts*. In both cases, the most significant feature is the preponderance of solid blue lines, which show the radio routes connecting nearly 1000 VOR stations across the country. Down low they're called "Victor Airways"; the name changes to "Jet Routes" above 18,000 feet.

Since the instrument system depends on up-to-the-minute navigational and communications frequencies, safe altitudes, and the like, IFR charts are updated regularly and frequently. They are quite satisfactory for VFR navigation, but be sure your radio-navigation skills are up to snuff—there's nothing to help you visually on an instrument chart.

En-route charts just won't cut the mustard when it's time to descend through the clouds for a landing; the instrument pilot must include *approach procedure charts* in his flight library. Published either in small bound volumes or as separate pages in loose-leaf binders, the "approach plates" provide guidance in all three dimensions of flight so that you will wind up directly over the airport, in most cases right in line with the landing runway. Small airports usually have just one procedure for finding the field from inside a cloud (you'd be surprised to know how many *very* small airfields are equipped to handle instrument traffic), while the metropolitan monsters may have a dozen or more to handle every conceivable condition of wind and weather.

To round out the collection of paperwork required for IFR operations, certain airports have standardized their instrument arrival and departure procedures: SIDs (Standard Instrument Departures) and STARs (Standard Terminal Arrival Routes).

NAVIGATIONAL COMPUTERS

They're the mousetraps of aviation—somebody is always coming up with a better way to figure time and distance problems, fuel consumption, and the correction for wind. But with the exception of the current wave of miniature electronic calculators, a navigational computer that your granddaddy used years ago will do just as good a job as a shiny new model. The advantage of the hand-held "electronic brain in a box" lies in the speed and accuracy of its calculations, with no writing or marking involved.

Nearly all the non-electronic computers that are available fall into one of two groups, based on the method of solving wind problems. One does it with a scaled slide that moves under a rotating plastic disc; the other is completely circular, with no sliding parts. But these computers are a lot like rowboats—turn them upside down and they all look alike— because the other side is nothing more than a circular slide rule with a few specialized functions for aerial navigators.

Used most frequently for time/distance/groundspeed problems, the slide-rule side deals with proportions. Line up the "10" indices and notice that the numbers on the two outermost scales are identical; for time and distance solutions, the outer ring of numbers always represents miles, the inner scale is al-

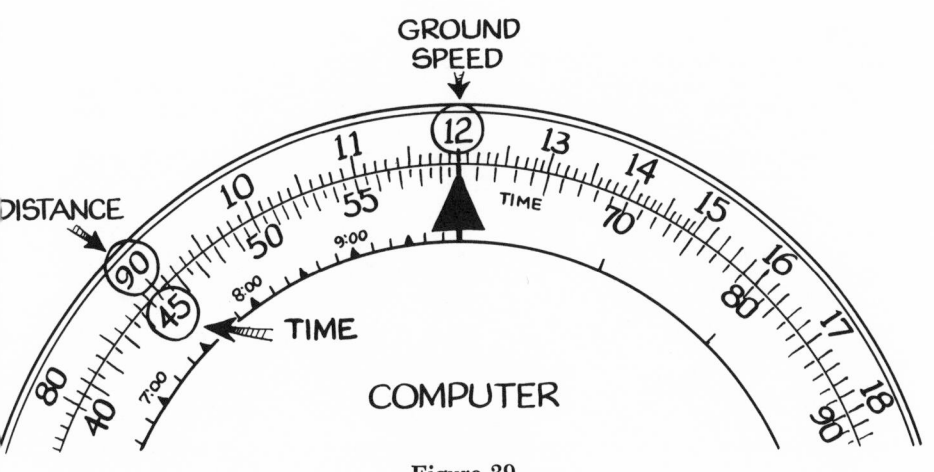

Figure 39

Slide-rule calculation

ways time. The prominent mark at "60" on the time scale is the *rate index* and is set opposite the number that represents miles per hour when it is known, or is the point at which you read miles per hour when rate is what you're solving for.

For example: Traveling across the ground at 120 miles per hour, how long will it take you to go 90 miles? Set the rate index at 120, which means that a full hour at that rate would move you 120 miles. The time required for only 90 miles has got to be something less than that—three quarters the time—so start at the rate index and move your pencil or fingertip clockwise on the distance (outer) scale to 90 miles and read 45 minutes.

It's always best to make an estimate of the time required (120 mph is 2 miles per minute) and of course you can come up with 45 minutes for that 90-mile trip without the computer —so why even take it out of its case for a problem as simple as that? But for those oddball numbers, the computer is a great help. For instance, your groundspeed is 134 mph—how long will it take you to travel 173 miles? At the outset, you know the answer must be in excess of 1 hour, so set the rate index on 134, move around to 173 miles, and read 77.5 minutes.

A more likely in-flight situation is the one in which you are not sure of your groundspeed, but you know how many miles have slipped under the wings in a certain period of time. This problem is obviously going to wind up with the groundspeed appearing at the rate index, so the only two figures left to work with are time and distance. Find the miles traveled on the outer scale, rotate the disc until the time flown is directly opposite miles, and read your groundspeed at the rate index.

The third case is the great anxiety reliever when you're concerned about getting home before the fuel tanks get down to the reserve level. When you know the groundspeed, it's easy to find out how long, therefore how far, you can travel. Just set the known groundspeed at the rate index and read the distance you can expect to achieve in the time remaining.

The same proportionate setups on the slide-rule side of your computer can be used for fuel-consumption problems by substituting pounds or gallons of gasoline for miles on the outer scale and pounds or gallons per hour for groundspeed at the rate index. It's just as important for you to form the habit of estimating before you compute; you should have a very good idea of what the answer should be, and let the computer do nothing more than confirm your suspicions.

While on the slide-rule side, there are several more functions you should know about. Most computers have a window (or two) in which you can line up the values of pressure altitude and outside air temperature for determining either true airspeed or density (performance) altitude. Once the altitude-temperature relationship is set up, true airspeed is read on the outermost scale directly opposite the number that represents the airspeed you see on the indicator; performance altitude (your computer no doubt is marked DENSITY ALTITUDE—same thing) shows up in another, smaller window elsewhere on the face of the computer. (You say there are two altitude-temperature windows on your computer? The "other one" is for determining true altitude, and will never be used in the course of normal flight operations. Forget it.)

There are always two properly spaced marks somewhere on the outer scale that represent the difference between nautical and statute miles. Simply line up the known value with the appropriate mark, which automatically translates that mileage into the other language at the corresponding point.

A direct-reading scale is printed on most computers to compare Celsius and Fahrenheit temperature readings, since weather observers insist on providing surface temperatures in degrees F. and temperatures aloft in degrees C.

THE WIND SIDE

The author of an old Irish toast must have had some aviator in him, for he included in his list of happy things: "and may the wind always be at your back." A constant tailwind—the stronger the better—would not only reduce the cost of flying, it would make your navigation chores a lot easier: Add the speed of the wind to the speed of the airplane, point the nose down the course line, and you will get to where you're going in a hurry.

A constant headwind (with which some pilots seem to be cursed) would be less pleasant, but would fit just as easily into the navigation equation: Airspeed minus wind speed equals groundspeed.

Direct headwinds and tailwinds exist only infrequently, and seldom for more than a very short while, since the movement of the air through which you're flying nearly always has some sideward component with relation to the intended course. When you have a forecast of wind direction and velocity, the problem becomes one of determining how much you'll need to turn the aircraft into the wind to stay on the desired course, and what the resulting groundspeed will be.

The brand name on your aeronautical "mousetrap" matters

not at all when you begin using the wind side. By one means or another, they all represent a scaled-down presentation of the *wind triangle*, which was used by the earliest aviators (whether they realized it or not) and is still used today by airline captains navigating jumbo jets around the world.

In its simplest form, the wind triangle is no triangle at all, but a line drawn from A to B representing the intended course. For convenience, assume the distance to be 100 nautical miles. An aircraft flying at a true airspeed (TAS) of 100 knots would make the trip in 1 hour; with no wind, there is no difference between airspeed and groundspeed. If a 30-knot headwind were forecast, a wind vector drawn to the same scale at the departure end would represent the distance and direction the airplane would be moved by the wind in 1 hour—if you placed yourself at Point A with zero airspeed, let the wind have its way for an hour, then turned off the wind and flew at an airspeed of 100 knots for an hour, you'd come up 30 miles short of Point B. To make it the rest of the way would require additional time, and it's obvious that the overall groundspeed would have been reduced by the velocity of the headwind. A 30-point tailwind vector drawn on the course line shows the opposite effect.

When the same 30-knot wind is blowing from the side, the wind triangle begins to take shape. Once again, wind effect is shown by drawing a wind vector at Point A, allowing the airplane to drift at the mercy of the wind for 1 hour, then resuming flight at 100 knots with no wind. You can aim for Point B, but you'll never make it. The wind has drifted your airplane off course, and the only way to complete the trip is by accepting additional flight time and the lower groundspeed that results.

Plotting a wind problem is cumbersome, the accuracy is subject to your skill as a draftsman, and it's a difficult job in the limited space of a small airplane's cabin. The navigational

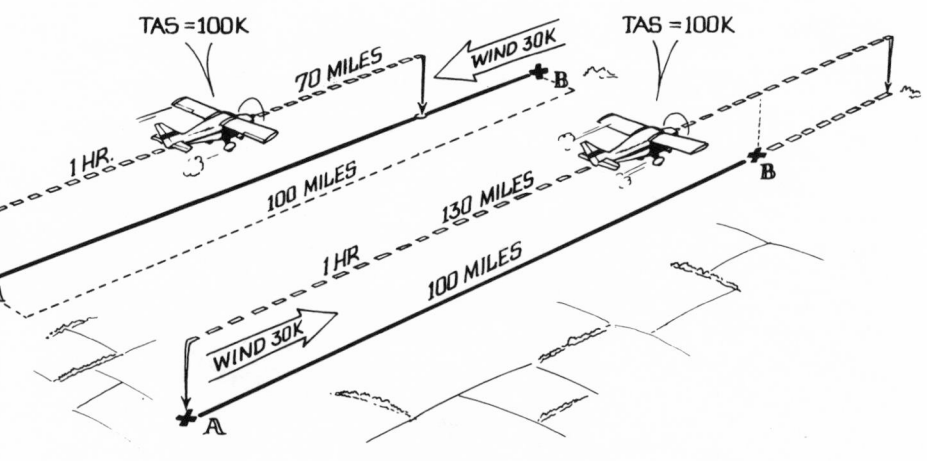

Figure 40

At a given true airspeed, the direction in which the wind is blowing will determine how far an airplane will travel in a specified time period. Be alert to the fact that regardless of groundspeed, the same amount of fuel will be consumed

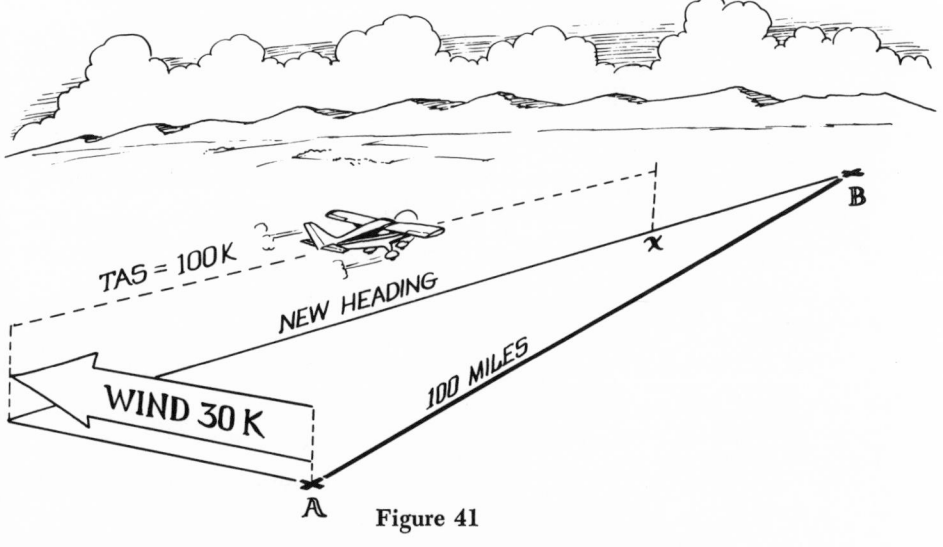

Figure 41

A direct crosswind, no matter how slight, will affect groundspeed; and the effect will always be negative

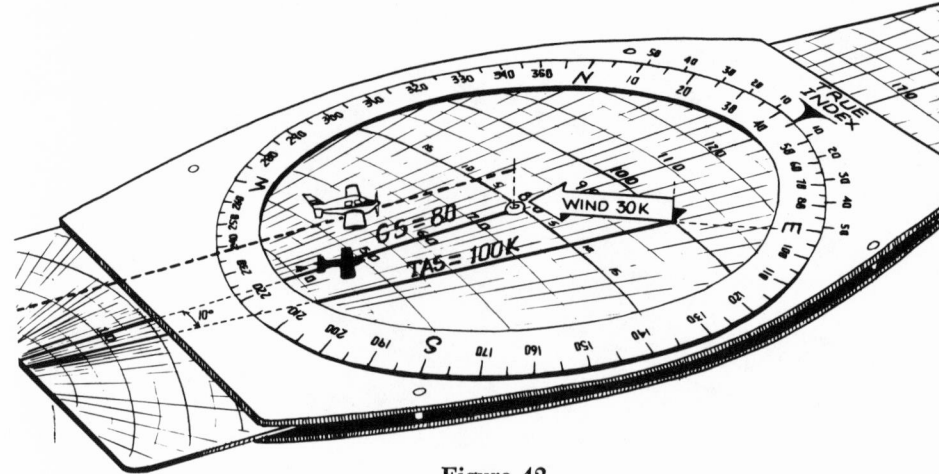

Figure 42

A navigational computer is merely a convenient way to
plot a wind triangle

computer provides a miniature plotting board and a rotatable
disc that reduces the wind problem to basic trigonometry.

Using a slide-type computer, the first input is the wind vector. Rotate the disc until the forecast wind direction is under
the top index, and make a pencil mark that represents the wind
speed, measured upward from the center of the disc. (The
vertical scale on the center line of the slide is uniform, so your
measurement can start on any convenient number.)

Next, rotate the ring until the course appears at the index.
Move the slide up or down to place the pencil mark (the tail of
the wind vector) on the line that represents true airspeed—
you've "drawn" a constant-airspeed triangle on the computer.

All that remains is to read the two answers: The wind-
correction angle in degrees left or right appears under your
pencil mark, and the groundspeed is read at the center of the
disc.

So much for estimating what things will be like based on the wind forecast. Would you like to know what the wind direction and velocity really are while you're flying? A simple reversal of the computer process provides the answer.

Previously, you knew the values of four navigational elements—wind direction, wind speed, course to be flown, true airspeed—and solved for the missing quantities of heading (course, plus or minus wind-correction angle) and groundspeed. In flight, assuming that you have done whatever is necessary to stay on course, you'll be certain of all those elements except wind direction and wind speed. Put the course at the top index, move the slide until groundspeed appears at the center of the disc, and make a pencil mark where true airspeed and wind-correction angle come together—you've just constructed the wind vector. Rotate the disc to put the pencil mark above center on the vertical scale and read the wind direction at the top index and velocity on the vertical scale. Considerably different from what the weatherman forecast for your trip? Don't be upset. This will be the case more often than not—forecasting wind is a tough job.

The circular, no-slide computers save space by using only the arrival end of the wind triangle, and deal with the concept of *wind components*. Insert the wind vector the same way as with a slide computer, rotate the disc to place the course at the top index, and the circular computer then provides the components of crosswind and headwind or tailwind. The outer scales relate true airspeed, crosswind component, and wind-correction angle by means of trigonometry, and groundspeed is derived by comparing the headwind/tailwind component to true airspeed. It's just a bit more involved, but a circular computer fits in your shirt pocket and has no parts to get lost.

ELECTRONIC NAVIGATION

The use of "black boxes" to help with aerial-navigation chores is growing by leaps and bounds. A modicum of skill in radio navigation is required of even private-pilot candidates, new facilities are added to the nation's system almost daily, and airborne electronic equipment has become less expensive in terms of cost-benefit ratio. You'll find some type of radio navigation gear in nearly all the aircraft flying today. Manufacturing and maintenance of electronic systems for aviation has generated a big, booming industry with its own name: *avionics*.

There are two distinct groups of radio-navigation aids available to the light-plane pilot—*low-frequency* systems, which transmit between 190 and 400 kilohertz (a "hertz" is 1 radio wave cycle per second) and the 108–118 megahertz *very-high-frequency* (VHF) band.

LOW-FREQUENCY RADIO NAVIGATION

Low-frequency radio beacons are old-fashioned—they are little-changed descendants of the very early radio aids—and should have been replaced by more sophisticated systems years ago . . . except that they won't go away. The transmitters are less costly to install and maintain than any other radio-navigation system, and the airborne equipment is equally attractive from a dollars-and-cents standpoint. As a result, few low-frequency beacons are decommissioned, and new ones continue to show up at small airports that have a need for an IFR capability but don't have enough instrument traffic to justify one of the more expensive (and more accurate) systems.

The broadcast pattern of a low-frequency transmitter accounts for its proper name, *non-directional beacon* (NDB), which indicates equal signal strength in all directions—you should expect a usable signal anywhere within the range of the

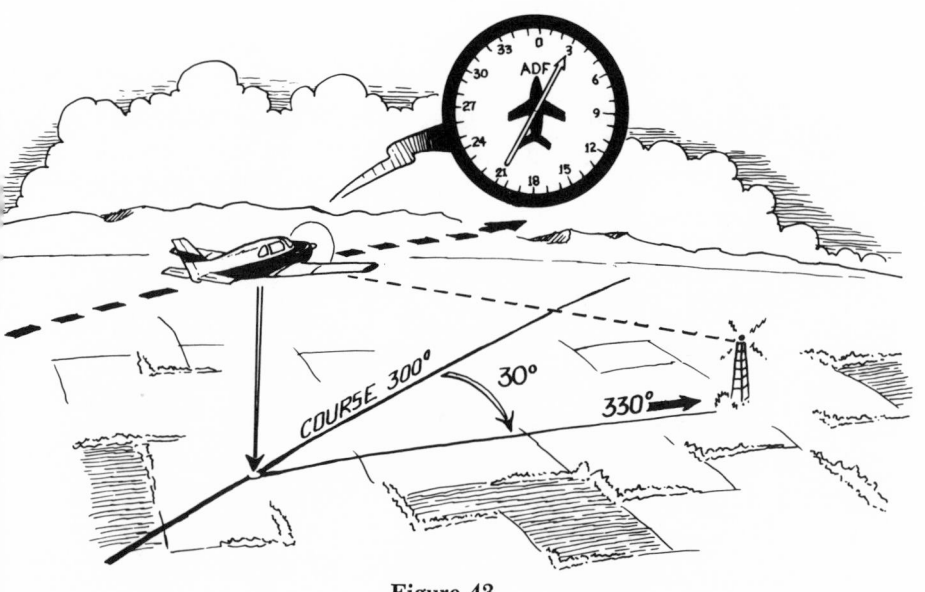

Figure 43

The Automatic Direction Finder (ADF) always points to the radio station and supplies the pilot with a relative bearing

station. (Certain other navigational transmitters are aimed at specific areas—such as down the center line of a runway—and are therefore *directional* signals.) The range of an NDB is determined by the amount of electrical power supplied to the transmitter, and while there are a few of these beacons capable of pushing a signal out for hundreds of miles, most of them are unusable beyond 25 or 30 miles. Today's NDBs are approach aids, for close-in navigation only.

When properly tuned (each NDB has its own two- or three-letter Morse-coded identifier), the low-frequency receiver in the airplane causes a needle to point to the radio station. Since, like a compass card, the pointer is superimposed on a full-circle dial marked off in degrees, it automatically finds the direction to the station—from which comes the airborne-equipment designation of *Automatic Direction Finder*, shortened to ADF. The dial doesn't move, so the number at the tip of the pointer represents the bearing (number of degrees measured clockwise from zero) of the radio beacon relative to the nose of the airplane. Most ADFs show an aircraft symbol on the dial to reinforce this indication.

To fly to the beacon, turn toward the pointer the number of degrees (left or right, whichever is shorter) it is displaced from zero, keep the needle on the nose, and sooner or later you'll fly over the transmitter. On a no-wind day, or with a direct headwind or tailwind, the flight path will be straight as a string. Under the influence of a crosswind, a drift correction must be applied to maintain course.

Once considered *the* method for long-distance point-to-point navigation (and still in vogue in many parts of the world because of its long-range capabilities), domestic use of NDBs is limited to instrument approach procedures and airport-finding for VFR pilots.

VERY-HIGH-FREQUENCY RADIO NAVIGATION

Before automatic direction finders came along, pilots often flew quite literally by ear. By listening to the tones in their headsets, they could determine their position in relation to one of the four courses, or legs, of a low-frequency radio range station. It was positive guidance, at least for those days, but four courses were hardly enough to handle the needs of a growing air transportation industry. In the 1950s, the *Very-High-Frequency Omnidirectional Radio Range* made its debut, and it is exactly what its name implies—a radio-navigation station broadcasting in the VHF band and providing course guidance in all directions. Someone had the good sense to shorten that handle to VOR (or "omni," as it's sometimes called).

There are approximately 1000 VOR stations in the United States, blanketing the country with electronic signals. Each has

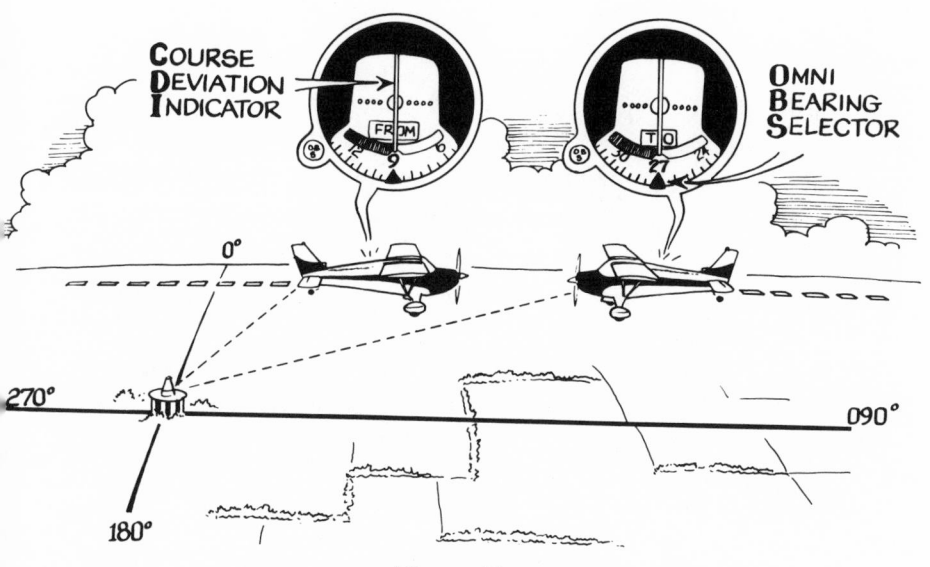

Figure 44

VOR-navigation instrument displays

a distinctive signature in Morse code, and some are equipped with a voice recording that announces the station's name between each series of dots and dashes: ". . . — . . . . — — — —/ Solberg VOR/ . . . — . . . . — — — —."

An omni station broadcasts an electronically phased signal, which is interpreted by a receiver in the airplane to provide information about the aircraft's position in relation to a pre-selected course. When speaking of VOR courses, the term "radial" is used universally and exclusively to indicate a magnetic course away from the station. Therefore, the 090 radial of any VOR is that course line that extends eastward from the transmitter. When traveling in that direction, you would characterize your position as "outbound on the 090 radial"; your magnetic course (and heading, if there's no wind to reckon with) would also be 090 degrees. Turn around and head back toward the radio station and you'd still be on the 090 radial, but inbound, and your magnetic course would be 270 degrees.

Each VOR transmitter is installed with the 360 (north)

radial lined up precisely with the magnetic pole, which provides nothing but magnetic courses to and from the station. Plan a cross-country trip using the radials from omni stations along the way, and you can forget about variation—everything is magnetic.

Airborne equipment for the VOR system of navigation comes in as many sizes and shapes as there are manufacturers, but every single one has three common features: an omni bearing selector (OBS), a course-deviation indicator (CDI or left-right needle), and a To-From indicator. The OBS references the entire presentation to a specific course, the To-From indicator tells a pilot what the result of flying the airplane in the direction he has selected will be, and the left-right needle indi-

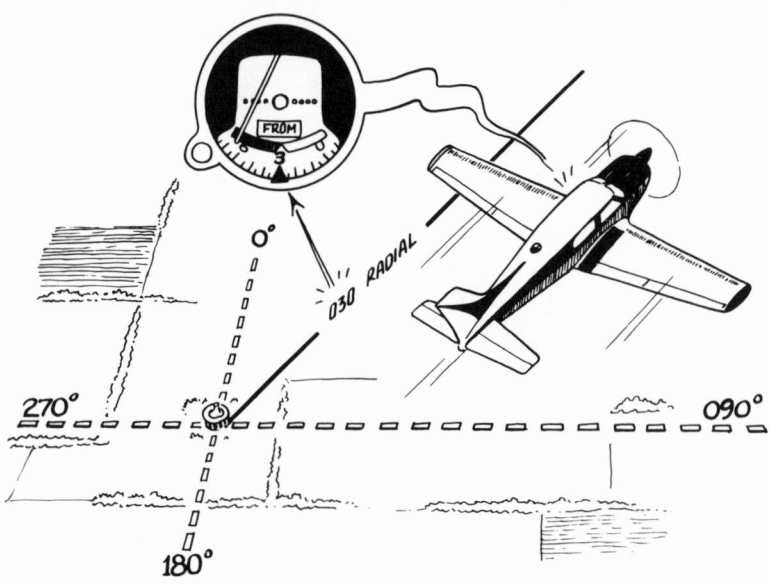

Figure 45

When the airplane is headed in the same direction as the course set in the OBS, the left-right needle will show the pilot which way to turn to get back on the desired radial

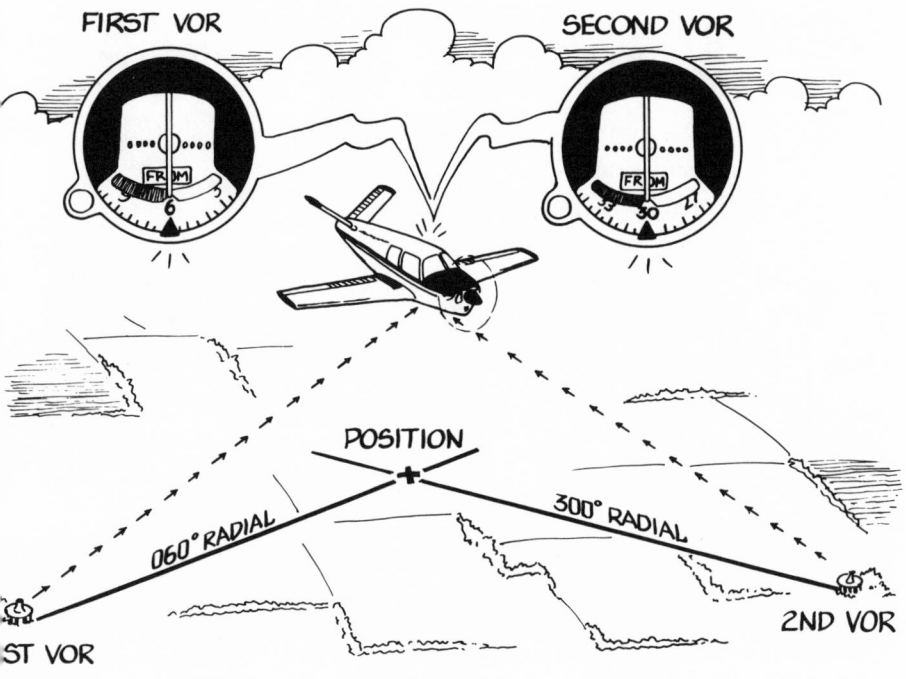

Figure 46

Determining position by plotting radials from two VOR transmitters

cates an on-course condition (needle centered) or left-right displacement.

Using these three tools, you can navigate very precisely to and from a VOR station on the radial of your choice or, as in the case of Victor Airways (the primary instrument routes across the country), the radials designated by the government. The VOR system also provides ample signals for solid cross-checks of your position on a VFR flight; when you can locate yourself on a specific radial of one omni station (rotate the OBS until the left-right needle centers with a From indication, then read the radial at the OBS index), repeat the process for a

---

second VOR, draw the appropriate radials on your chart and where the lines cross is your position—or *was* a couple of minutes ago.

The VOR system has a major drawback, one shared by all VHF transmissions—the reception distance is limited by line of sight. Unlike low-frequency signals, very-high-frequency radio waves don't follow the curvature of the Earth, but tend to travel in straight lines. Get behind a mountain or on the other side of the horizon and you'll lose contact with a VHF transmitter—if you can't see it, you won't receive it. In general, the secret to stretching the distance for VHF reception is altitude, shown by this table:

AIRCRAFT ALTITUDE (FEET)	RANGE (NAUTICAL MILES) OVER FLAT TERRAIN
500	28
1,000	39
1,500	48
2,000	55
3,000	69
5,000	87
10,000	122
15,000	152
20,000	174

Most domestic VORs incorporate distance-measuring equipment (DME), a feature that provides a cockpit readout of nautical miles between airplane and radio station. The black box in the airplane (a unit completely separate from the VOR receiver) interrogates the ground station, measures the time required for the reply, and converts the time lapse into nautical miles. You can now determine your position very accurately with the information from just one station—a radial from the

VOR portion, and distance from the DME. Once the time lapse is known to the airborne equipment, it's a simple matter for electronic circuitry to add a groundspeed readout (based on the rate of change in time lapse when you're flying toward or away from a VOR/DME), and just as easy to convert *that* into time-to-station—and most of the current crop of DME units have those features as standard equipment.

THE INSTRUMENT LANDING SYSTEM

Approach-procedure charts describe specific electronic routes to a point from which a pilot should be able to see a runway and land, even under conditions of very low visibility. The electronic routes can be VOR radials, or courses based on the signal from a nondirectional beacon. Distance information can be provided by intersecting radials, DME readouts, or, in the simplest situation, timing based on airspeed and the pilot's best guess of wind direction and velocity.

Navigation with such tools is accurate enough to permit safe approaches to within 400–500 feet of the ground, but these two-dimensional procedures have a common shortcoming— there's no vertical guidance, so the pilot descends to the specified minimum altitude at whatever rate he considers proper.

Enter the Instrument Landing System (ILS), a three-part radio installation that provides super-accurate guidance in left-right, vertical, and distance dimensions. The heart of the system is the *localizer*, an electronic extension of the runway center line. This very narrow beam of radio energy is processed and displayed as steering commands on the VOR indicator in the airplane—but when tuned to a localizer frequency, the left-right needle becomes four times as sensitive, and therefore four times more accurate than it was when tuned to a VOR station.

Vertical guidance comes from a *glide-slope signal*, which drives yet another bar or pointer on the face of a modified VOR

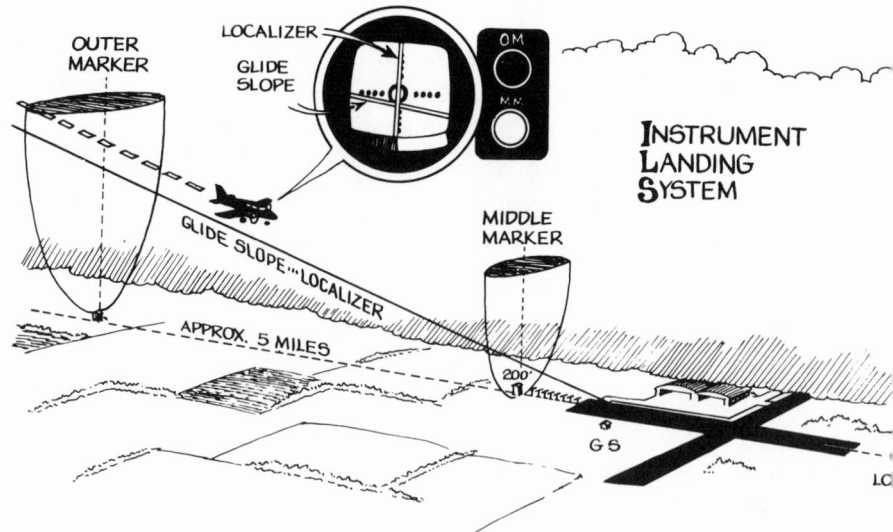

OUTER
MARKER

LOCALIZER

GLIDE
SLOPE

OM

MM

GLIDE SLOPE

LOCALIZER

APPROX. 5 MILES

MIDDLE
MARKER

INSTRUMENT
LANDING
SYSTEM

200'

GS

LO

Figure 47

Instrument Landing System

display, and provides a smooth, shallow electronic glide path to the runway. Two or more *marker beacons* are installed along the center line of the approach path, and trigger audiovisual signals for distance information.

With the three-dimensional guidance thus available, an instrument pilot can descend to a point very close to the landing area—typically 200 feet above the runway, and right on the center line if he's kept the localizer needle centered.

RADAR NAVIGATION

Now firmly established as one of aviation's household words, *radar* is a manufactured term, an acronym that expands to *Radio Detection And Ranging*. It is a means by which electrical energy bouncing off an aircraft is returned to the transmitter and processed into an image on a cathode-ray tube. An observer can determine distance and direction, issue instruc-

---

tions to the pilot to cause the image to move across the scope in a desired direction, and therefore accomplish the navigational chore entirely from the ground. This is the very easiest type of navigation, because the pilot merely does what he is told.

The fly in the otherwise ideal ointment is the fact that radar navigation requires one-to-one pilot-controller communication, which, if interrupted, could result in complete chaos in the airspace system—to say nothing of the oversaturation of available communications channels. For these reasons, radar has evolved into the full-time *monitor* of aircraft whose pilots are navigating by other means. There just isn't enough time to talk to everybody.

But get yourself into a spot of navigational difficulty—disoriented, unsure of your position, or just plain lost—and a call to the nearest radar facility (the network blankets all but the most remote parts of the country) will provide all the navigational assistance you need to find your way to an airport, or at least get you headed in the right direction. Get yourself into *real* trouble—lost in weather conditions you're not prepared to cope with—and the controllers will take you to their electronic bosom and guide you all the way to a runway if necessary.

In the beginning (the British developed radar techniques to a fare-thee-well during World War II, when they tracked Luftwaffe fighters and provided vectors—headings and altitudes—to RAF pilots for the intercept), radar depended on nothing more than the echo of transmitted electrical energy, which meant that larger, closer, more metallic targets afforded better images than distant, small, perhaps fabric-covered airplanes. Adding to the quality and validity problems were mountaintops, precipitation, and even occasional flocks of birds, which also echoed the radar beams and did nothing for dependable interpretation of the images on the scope. Today, all commercial aircraft and a constantly increasing number of the general aviation fleet are equipped with *transponders*—

miniature coding devices that return the radar signal through a computer. Instead of dealing with the confusing, variable-strength returns of early radar, the controller works with positive data about aircraft position, and in most cases ground-speed, altitude, and projected flight path as well. Needless to say, radar has become the sharpest tool of air traffic control.

VHF DIRECTION-FINDING SYSTEMS

As a last resort, the pilot whose fountain of navigational knowledge has run completely dry can take advantage of the direction-finding equipment installed in most Flight Service Stations and control towers. When the situation is made known to a ground station, the controller will have you hold down the microphone switch for ten seconds or so, which activates the direction finder and indicates to him the bearing from station to aircraft. Successive bearings (or cross-bearings obtained from other controllers who are simultaneously fixing your position as you transmit) confirm your position and movement, and make it possible to give you a heading or "steer" to bring you directly over the airport.

All this is accomplished using normal VHF communications frequencies, and makes emergency help available to VFR pilots (yes, "lost" is an emergency, even in good weather conditions) without disrupting the primary IFR-monitoring responsibilities of the radar controllers.

AREA NAVIGATION

Just about any spot in these United States can be located on a radial of at least one VOR, but it's difficult to go directly to that spot without first getting on the radial. The addition of DME equipment would make it possible to fly around an arc of the proper radius until intercepting the desired radial, but that would be time-consuming and inefficient.

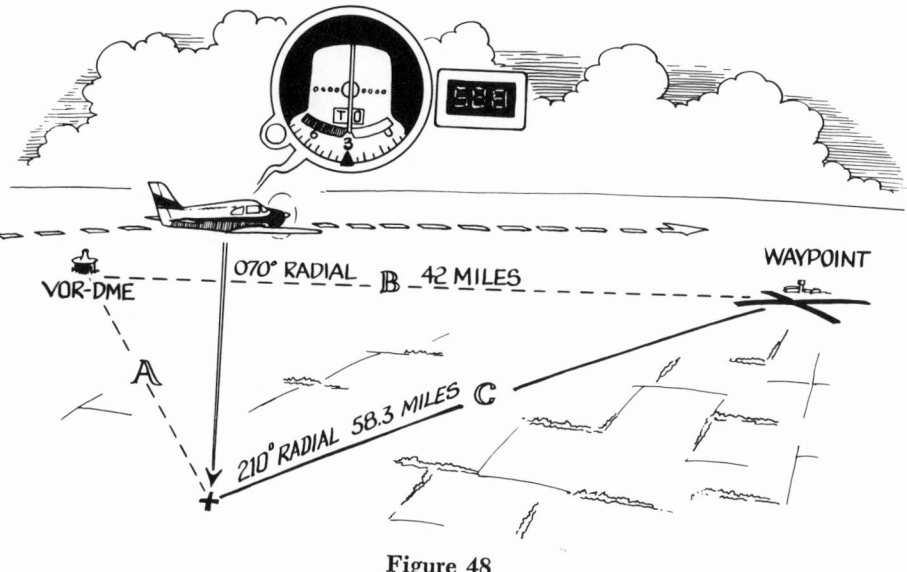

Figure 48

Area navigation (RNAV) is a continuous electronic solution of a simple trigonometric problem

A clever application of space-age miniaturization of electronic components has developed a new concept called *area navigation,* which makes it possible to fly directly to any point within the signal area of a VOR-DME station. An on-board computer (usually about the size of a lunch box) uses the radial-distance information from the station (side "A" of the triangle) and pilot-supplied data about the destination, or *waypoint,* as it's known in RNAV language (side "B" of the triangle), to constantly update the side "C" of the triangle. The result is displayed as course information on the VOR left-right needle, and distance-to-the-waypoint information in digital form.

Area navigation permits straight-line flight to any point the pilot chooses, with no need to fly from one VOR to the next. In effect, through the magic of electronics you can move a VOR-DME station to literally any place you'd like it to be—subject to line-of-sight limitations, of course. It's a few extra bucks, but a lot of extra utility and efficiency.

If history is any indication of what's to come, general aviation pilots can expect to share in the technological advances that have provided airline and military operators with some rather heady navigational devices. Initial development and application usually take place where the money is; then the fancy black boxes filter down to the less affluent users.

There's no doubt that we'll eventually be guided through the skies with signals from *navigational satellites,* and there are already a number of *ultra-low-frequency transmitters* in operation to provide extremely accurate position fixes anywhere on the face of the earth. In *inertial navigation,* sensitive accelerometers (delicate gyroscopes mounted in all three planes of aircraft movement) combine with sophisticated computers to determine where they've been, where they're going, when they'll be there, and a host of other data, with incredible accuracy.

But there's one navigation system that will never be replaced —the man who gets the whole thing going, who supplies the information so the black boxes can make their magic, and who has the ability to make judgments about the progress of the flight. Old-time aerial navigators flew from one known position to the next, often keeping a finger pressed firmly to a landmark on the chart until they spotted the next one. They were developing what is surely the most important quality of a good navigator, regardless of the complexity of his equipment: He knows where he's been, where he's going, and when he expects to be there.

Part Three

BOUNDARY LINES

The Airspace
Rules of the Air
All About Airports

The Airspace

Ah, the freedom of flight, shared by the birds and certain privileged humans: no lane markers, stoplights, interchanges, or traffic cops, nothing to limit your activities but the capabilities of man and machine.

You can believe that, if you choose, but it's a cavalier attitude that just won't hold up in today's airspace. There was a time when there weren't enough flying machines to require limits on their activities, and nobody really cared one way or the other. Airplanes were novelties, and if they ran together now and then, the pilots should have known better. Now, however, although there are still parts of our country sparsely populated enough to allow relatively unrestrained flight, most of the sky between our borders and coastlines is under some kind of control. Birds cannot be punished for violations, so their freedom of flying is absolute, but human pilots must accept certain restraints if we're to wind up with an orderly, safe sharing of this great national resource we fly in.

Political philosophies notwithstanding, government agencies are set up to do for citizens what the citizens can't do for themselves. One of these things is to devise and maintain a national airspace system that can keep the users happy and safe. This is an impossible task—some users would never be happy unless they had the whole sky to themselves, and there's no way to guarantee safety so long as people are flying the airplanes—but the airspace designers do their best, and the system does a pretty good job so long as pilots understand their privileges and limitations while operating within that airspace.

As the complexion of aviation in the United States changes (there are more people doing different things with different types of airplanes every day), the airspace dimensions must be redrafted now and then or fall hopelessly behind the needs of the users. It's assumed, though, that certain basic divisions will survive indefinitely, and these are the substance of what follows. However, *please be forewarned that this material is not timeless.* There is no substitute for staying on top of the rules, and no excuse for not doing so.

FIRST, THE BIG PICTURE

Visibility: That's the key. When the air is clear enough for pilots to *see* each other, they can be relied upon to *avoid* each other. When the visibility drops to the point at which a pilot can't expect to see the other guy in time to avoid a collision, someone else must shoulder the responsibility. A long time ago, the powers that were (and still are) decided that 3 miles would be a good distance to use, considering the probable closing speeds of aircraft and human reaction time. The same powers decreed that the fellows on the other side of the microphone, the air traffic controllers, should be in charge of the airspace when pilots couldn't see for themselves. Boundaries had to be drawn so that controllers and pilots alike would know who was responsible for what, which led to the legal partitioning of the atmosphere into *controlled* and *uncontrolled airspace.* It's a very simple and logical division. In general, with visibility 3 miles or more, pilots are on their own, but when the atmosphere begins to clank up with smoke, clouds, haze, rain, snow, or whatever and the visibility drops below 3 miles, air traffic controllers assume the responsibility of keeping airplanes under their control from running into one another.

At once, communication between pilots flying in poor visibility and the controllers who separate them becomes imperative, as does the requirement for those pilots to be operating in accordance with a clearance so the controllers can know where everybody is. With that in mind, the first and foremost rule of airspace discipline is: When inflight visibility drops below 3 miles, the only people allowed to fly in controlled airspace are those with a clearance to be there. Above 10,000 feet MSL, the visibility must be 5 miles—there are more, faster-moving airplanes up there—but in general there are few restrictions on flight so long as pilots can see.

In the early stages of learning how to fly, you needn't be concerned about stepping on anyone's toes at the very high altitudes; from 18,000 to 60,000 feet, the airspace over the entire country is controlled 24 hours a day regardless of visibility, and the requirements for flying up there preclude all but instrument-rated pilots in rather well-equipped aircraft—that's Positive Control Airspace. A layer between 14,500 and 18,000 feet is also constantly controlled, but there are no restrictions on visual flight when visibility is 5 miles or more.

FOR NOW, LEAVE HIGH FLIGHT
TO THE BIG GUYS

Except for crossing that wrinkle in our country's hide out west and making an occasional higher excursion to top a cloud layer or get a kick in the tail from a robust breeze, you will probably keep your light-plane operations in the lower levels of the atmosphere, those altitudes between the surface and 10,000

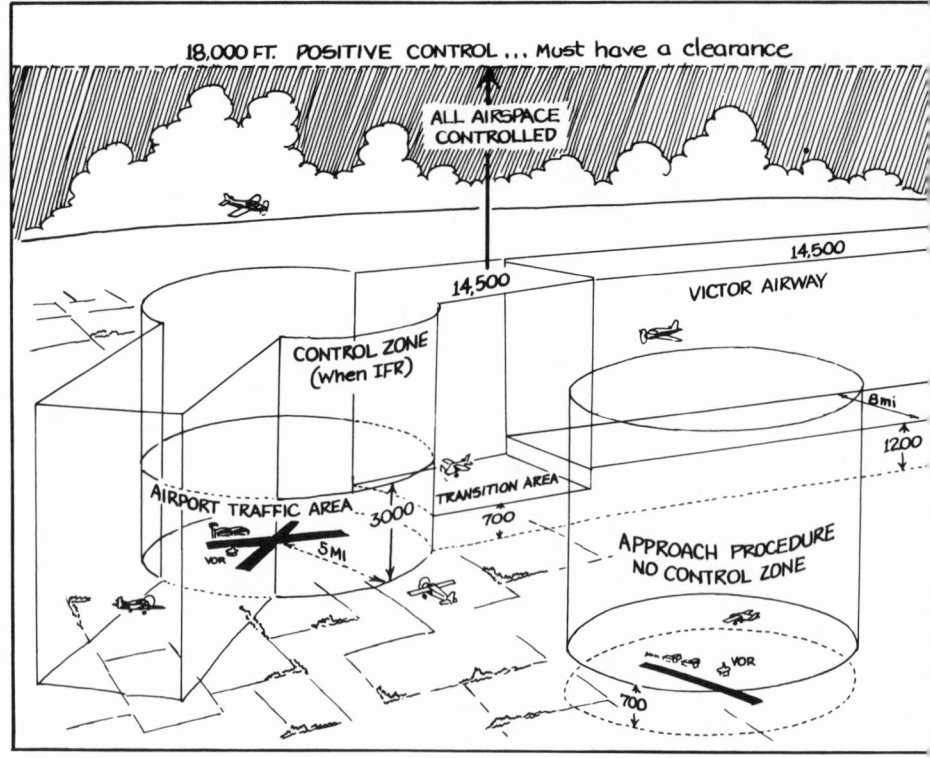

Figure 49

Typical airspace restrictions

feet. Assuming that you'll not be out flying around in visibilities of less than 3 miles, that surface-to-10,000 foot slice of airspace is all yours—the "high seas" of aviation, if you will. Even the airways, those invisible electronic paths from VOR to VOR, have no sanctity, so long as the visibility holds up, but to guarantee protection for the folks using them when Instrument Flight Rules (IFR) are in effect, airways are insulated by vast chunks of sky 4 nautical miles on either side of the centerline and extending from 14,500 feet MSL down to 1200 feet above the ground. Most of the airspace east of the Mississippi is so designated, because the number of VORs and associated airways produces an almost complete overlap of all those 8-mile strips. In the less populated and therefore less traveled parts of the U.S., you'll find airways widely separated, with large open spaces in between. When that situation prevails, uncontrolled airspace rises unhindered from the surface of the Earth to 14,500 feet above sea level.

(It's possible and perfectly legal to aviate in the lower levels of uncontrolled airspace when the visibility is as little as 1 mile, but pilots of even the slowest airplanes find that mountains and TV towers and other airplane-bending obstacles grow very rapidly out of the 1-mile murk. Flight under such conditions is best left to flyers with lots of experience, and for those who qualify, why not go IFR? The safety advantages of proceeding under instrument flight rules and the attendant umbrella of protection offered by ATC far outweigh the occasional inconvenience of obtaining a clearance.)

The simple division of the airspace into controlled and uncontrolled areas begins to break down in the vicinity of airports, where the increased volume of traffic and the mix of aircraft types—meaning vastly different speeds, sizes, and performance characteristics—require additional restraints for adequate traffic control. When the weather is good and aircraft are approaching from all directions, a tower controller should have some working room, a piece of airspace set aside for those flights operating to and from his airport. His needs are met by the Airport Traffic Area, a cylinder of airspace 5 miles in radius and 3000 feet deep, centered on the airport. Since the only reason for being in an Airport Traffic Area is to land or take off, pilots are required to be in radio communication with the tower whenever they're in that airspace. There's also a speed limit—150 knots maximum will keep you out of trouble. The ever-present see-and-avoid rule should keep your head swiveling even more than usual within 5 miles of an airport. Since it is designed to help the controller to do his work better, the Airport Traffic Area disappears when a part-time tower closes

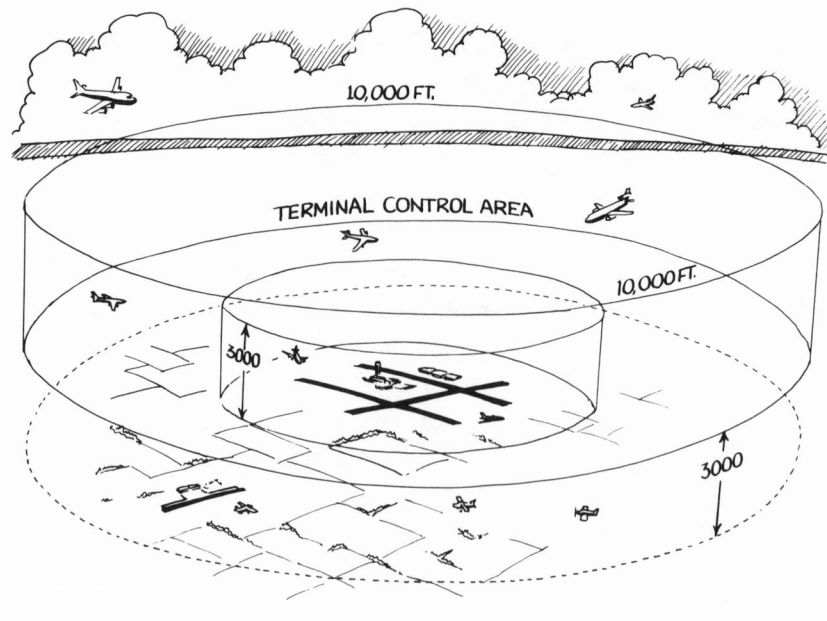

Figure 50

The Terminal Control Area; an air traffic control concept
designed with the safety and operational interests of all
aviation users in mind

shop for the day (you can expect to encounter an increasing
number of less-than-24-hour tower operations; the government
is slowly coming around to the realization that pilots at low-
volume airports can handle their own traffic problems in the
wee hours of the night), but a closed tower is no reason to relax
your scan for other airplanes. Keep your eyes open, monitor the
tower frequency, and always expect that somebody else will be
trying to use the airport at the same time you're there.

The Airport Traffic Area just won't cut the mustard at certain very busy terminals—5 miles isn't broad enough, 3000 feet
isn't deep enough—so even more restrictive airspace has been
established: the Terminal Control Area. A low-level analog to
the Positive Control Airspace above 18,000 feet, the TCA provides more elbow room for controllers and more airspace to

contain the comings and goings of a *lot* of high-speed, high-performance aircraft.

The overall dimensions of a TCA vary from one area to the next, but most of them start at the surface and extend upward as high as 7000, 8000, even 10,000 feet. It's not as bad as it sounds, because the underside of this airspace is stepped. The surface-to-top portion may extend outward for only 5 miles or so—that takes care of the close-in traffic—and the floor of the next layer may be at 2000 or 3000 feet above the ground, allowing low and slow aircraft to go safely about their business underneath the TCA. Two or three such layers (sometimes more, if they're needed) make a Terminal Control Area look very much like an upside-down wedding cake, with the busiest airport where you'd expect the bride and groom to be.

Airspace designers and aviation rule-writers have come to the conclusion that TCAs won't accomplish the full measure of their intent unless pilots are required to do more than just be aware of their presence, so certain additional requirements are levied on flights operating therein. Not only must pilot-controller communications be established but a clearance must be received to enter the TCA. To fly into or through or out of the wedding-cake around one of the really big terminals (New York, Chicago, Los Angeles, and the like—there are fewer than a dozen of these Group I TCAs), your airplane must be equipped with a radar transponder for aiding positive identification by the controllers. Sensibly, student pilots on solo flights are not permitted to fly in a Group I TCA. Less-busy airports (such as Cleveland, St. Louis, Seattle, Las Vegas, Kansas City) are protected by Group II Terminal Control Areas, which have requirements that are a bit less stringent. In almost every case where small general aviation airports lie close to the major terminal and would be severely hampered by the restrictions of the TCA, low-level notches or corridors are cut into the cake to allow access for the little guys.

When visibility deteriorates to less than 3 miles (or when the lowest solid layer of clouds—the "ceiling"—is less than 1000 feet above the surface), instrument flight rules take effect and turn *all* controlled airspace into the official property of Air Traffic Control; an instrument clearance is an absolute necessity if the controllers are to fulfill their responsibility for aircraft separation. Non-instrument pilots are automatically excluded under these conditions.

The 1200-foot floor of controlled airspace undulates with the contours of the terrain and overlies most of the country. It is studded here and there with Airport Traffic Areas and TCAs. But in IFR weather, ATC must have jurisdiction over all the airspace that would conceivably be used by pilots following published instrument-approach procedures. To accomplish this, nearly all airports that are accessible to instrument pilots are embedded in Control Zones; these zones begin at the surface, are usually 5 miles in radius, with extensions necessary to accommodate approach paths (the extensions often give a Control Zone the appearance of a keyhole), and rise to 14,500 MSL.

Before the government will provide a Control Zone, thereby assuming separation responsibility, a communications link between the airport and the ATC must be established. For the hundreds of airports that qualify for a published instrument-approach procedure but lack enough air traffic to justify a communications facility, a Transition Area is designated. Just as the name implies, a Transition Area protects instrument pilots moving from controlled to uncontrolled airspace during an approach procedure, and vice versa during an instrument departure. The 1200-foot floor of controlled airspace is lowered to 700 feet above the ground in a Transition Area. In effect, it's a Control Zone that doesn't quite reach the ground.

Certain parcels of airspace are set aside for special uses; the restrictions range from absolute exclusion to "friendly" warnings. For example, the airspace above the Capitol building, the White House, and several other locations where national security is a concern are designated *Prohibited Areas*, to be avoided at all times.

With published hours, altitudes, and well-charted geographic limits, Restricted Areas may contain artillery ranges, military aircraft conducting air-to-air refueling, or other such operations that require exclusive use of a chunk of the sky.

The military air services also confine their pilot training to designated areas surrounding their training bases; these areas are wisely avoided or flown through with a little altitude and a lot of looking because of the high speeds and small profiles of most military jet trainers. The vertical dimensions and times of use are prominently displayed on aeronautical charts.

Any military operations area (MOA) should get and hold your attention—it's the official way of saying "You may be sharing this airspace with a lot of other folks whose duties require more eyes-in-the-cockpit time than usual. Since you have nothing else to do, *look around!*"

Of all the special-use airspace, perhaps the most hazardous to light planes is also the type you're least likely to recognize: Low Altitude Training Routes, which are used by military pilots to perfect their skills in navigating close to the ground (as low as 500 feet) at very high speeds (up to 500 knots). Laid out over sparsely populated areas wherever possible, they're called Olive Branch routes, and you'll find their locations displayed in Flight Service Stations and most airport operators' offices in those areas directly affected. The best procedure for the light-plane pilot is to know where the Olive Branch routes are and to *stay away*. The reaction time required to prevent a collision at a closing speed of 500 knots or more is considerably less than the time available.

Air Traffic Controllers will not clear instrument flights through airspace that has been previously reserved for military exercises; that keeps the IFR pilots honest, but flying through one of these Military Operations Areas under *visual* rules puts you entirely on your own in regard to traffic separation. MOAs are indicated on all aeronautical charts, and it's best to avoid them during the hours of intended use.

ROOM FOR EVERYBODY

The best way to stay out of trouble (besides the obvious legal implication, the unjustified use of someone else's airspace often results in broken airplanes) is to stay on top of changes in the designation of airspace and the procedures for its use. As in everything else pertaining to flying, education is the key, because a smarter pilot is almost always a safer pilot.

Rules of the Air

Long before the federal government took over the responsibility for writing and enforcing the rules of the air, pilots had developed regulations of their own. These were unwritten laws, based on common sense and a common desire to terminate each aerial adventure at a location and in a manner completely under the control of the pilot. So long as no one could be charged with responsibility when a flying machine and perhaps its operator met an untimely end, there was no need to document any limits or restrictions.

When government got into the act and required reports whenever an accident took place, the rules had to be put on paper. In the United States Airmail Service, an ill-fated federal adventure of the 1920s that killed nearly all its pilots before it was abandoned, those in authority found it necessary to remind pilots officially of such elementary safety practices as "Don't take your machine into the air unless you are satisfied it will fly" and "Never leave the ground with the motor leaking." It's a good bet that those two rock-solid and timeless rules were written into the aviation books shortly after some air-mail pilot had a crackup in a machine that he *thought* would fly or that had a leaky motor, or both.

Once the precedent had been established, whenever a pilot figured out some new way to get into trouble, a rule was added to discourage repeat performances. The rules of flying, given the full force of federal law when Congress empowered the Federal Aviation Administration to write and enforce them, grew in number and scope, so that today only the most ingenious pilots have aircraft accidents outside the law.

The Federal Aviation Regulations (FARs to most pilots) are all-encompassing—from the greenest pilot in a little airplane to the airline captain in a jumbo jet and everything in between—and they represent the government's guidelines for discharging its responsibility to the aviation community. The FARs may appear unduly restrictive at times, but they are an attempt to provide the safest possible environment in which aviation interests at *all* levels can operate. There may be some regulations that are not grounded in flight-safety considerations, but they are few and very far between.

Rules change every now and then, but it's not often that an existing regulation is dropped. There are, however, frequent revisions and changes of emphasis of which every pilot must be aware. There's a responsibility to stay current, to know what *today's* rules are. The best way to accomplish this is to subscribe to one of the commercial regulation services, from which you'll receive a basic book of regs and changes as they become effective. At the very least, a serious pilot should spend some time with his instructor every once in a while to find out what's new. Ignorance of the law is no more an excuse in an aviation trial than it is in traffic court.

The discussion that follows is very general, and deals with regulations that have been on the books for many years and are likely to remain in force for a lot more. For the most part, they are the rules designed to keep airplanes from running into each other—a midair collision is the ultimate sin in aviation—and into things other than the runway at the proper time and place. This is necessarily one person's opinion of just a few of the host of regulations for which a pilot is responsible. Not a single one of them is immune to changes in content or interpretation. Whatever else you take from this chapter, resolve to get your hands on a set of the rules that affect *your* flying and to make yourself aware of the *now* situation; there's no substitute for current legality.

GENERAL OPERATING RULES

Part 91 of the federal regulations governs your actions as a pilot in any type of aircraft, from balloons to Boeings. Even airline captains, whose comings and goings are covered more specifically in other sections of the rulebooks, must ultimately return to Part 91 for general guidance. It's the jumping-off point for all flight operations.

This keystone of aviation regulations establishes the pilot in command as the one directly responsible for whatever happens during the course of a flight—and since he can't be expected to shoulder total responsibility without the legal power to make his own decisions, each pilot is granted total authority regarding the operation of his aircraft. When confronted with an emergency that requires action *right now,* he is required to do whatever his training and judgment indicate is the safest thing to do. If such action results in a violation of the rules, so be it; he may be asked to justify his procedures after it's all over, but a report is a small price to pay for avoiding an accident.

Before beginning a flight, the pilot must be familiar with all the information that is available concerning that flight; this is a big order but one that must nevertheless be filled. Specifically, your interest should center on the wind, weather, fuel supply, and alternate courses of action to take if the flight doesn't proceed as planned, but the rule is very general and may embrace local practices such as traffic-pattern directions, noise-abatement procedures, ad infinitum.

The documents that establish the ownership and airworthiness of the airplane must be up to date (current registration papers and a certificate of airworthiness must be on board the ship, and the pilot is responsible for their presence and validity whether the airplane is owner-flown or rented), and there are regulatory concerns for the pilot's condition as well. You may not fly an airplane for eight hours after taking a drink—that's eight hours between bottle and throttle no matter how small the drink—and it is against the law to act as pilot in command any time you are aware of a medical problem that would prevent you from passing a flight physical. In other words, before every flight you become your own medical examiner.

While you've got your doctor hat on, be certain that the only trip your passengers intend to take is in the airplane—you may not take aloft anyone obviously under the influence of alcohol or drugs. No blue-nose statutes these, but sensible rules that take into account the remarkable increase in the effects of alcohol and other drugs with altitude, and the safety problems that might result.

Some automobiles are rigged so that they won't start until the seat belts are fastened; we haven't progressed (?) to that point in the airplane business yet, but the regulations are very clear on the use of human restraint systems. Everyone on board must be wearing a seat belt during takeoff and landing, and the pilot must make certain that all his passengers have been so notified. The pilot must wear his belt throughout each flight.

GENERAL FLIGHT RULES

Flight safety jumps right off the pages when you get to this section of the Federal Aviation Regulations. At the outset, you are reminded of the illegality of flying so close to another airplane that a collision hazard is created—which doesn't prohibit formation flying, only that which is done without the express knowledge and consent of all the pilots involved.

Realizing that formation flight is not very popular (or necessary) among civilian aviators, it's important to know exactly how to resolve the conflict when your airplane and another appear headed for the same spot in the sky. Perhaps one of the earliest rules of flight (the airmail-service pilot manual stated, "If you see another machine near you, get out of its way"), the "see and avoid" principle, is also the most fundamental. Simply stated, whenever a pilot can *see* another aircraft, he will *avoid* it. And the regulations once again put the responsibility right where it belongs by requiring all pilots to maintain vigilance so that they *can* see their fellow fliers when weather conditions permit.

Once spotted, another airplane is to be avoided through the use of a logical set of procedures that provide mutually acceptable actions to prevent a collision. At the top of the list of priorities is an aircraft in distress; that pilot has enough to worry about, so he has right-of-way over anything else in the air. In more likely situations, maneuverability has most to do with who is supposed to get out of the way, so balloons, which are at the complete mercy of the wind, take precedence over other aircraft; gliders have the right-of-way over airplanes, airships, or helicopters; blimps (airships) will be avoided by all other powered aircraft. (Ever try to make a sharp turn in a blimp?)

In general, when there are two airplanes at the same altitude on a converging course, whichever is to the other's *right* has the right-of-way. Should the two be approaching head-on, both should alter course to the right. When one aircraft overtakes another, the faster flyer is expected to move out to the right and pass well clear of the other ship. Finally, with respect to the pilot's increased attention to the job at hand, an aircraft on final approach to land or one in the actual landing process has right-of-way over others in flight and on the ground. If two aircraft are approaching the same airport for the purpose of landing there, the one at the lower altitude takes precedence; it's a first-come, first-served situation.

The see-and-avoid rules are easy to learn and even easier to use. If the aviator's Utopia is ever achieved, if we ever get to the point where each and every pilot follows the right-of-way rules to the letter, the midair collision will be a thing of the past.

Having provided a legal solution to the problem of airplanes running together, the rule writers turned their attention to minimum safe altitudes and created another series of "thou shalt nots," to prevent airplanes from crashing into the ground or other obstacles. As a general rule, you must always maintain enough altitude to allow an emergency landing in the event of power failure. In the case of a single-engine airplane, a power failure certainly reduces the options, but you must fly high enough to glide to a spot where you can put 'er down without undue hazard to the people and property below. In other words, you should always fly from one potential forced-landing site to the next.

You may fly as close as you like to the surface of open water or unpopulated land, but never closer than 500 feet to any person, vehicle, or structure. Over populated areas—cities, towns, groups of people—you must take into consideration the highest obstacle within 2000 feet of your airplane at any time and maintain an altitude 1000 feet above that, to protect the public.

It seems that aviators are sometimes afflicted with the morbid curiosity demonstrated by drivers when an accident takes place on the highway. In recognition of the increased midair-collision hazard and possible interference with relief activities that would be created by a swarm of aircraft, pilots should anticipate that natural disasters, air shows, major fires, and the like will be off limits. To put teeth into the common-sense rule that should keep you away, the FAA will apply temporary flight restrictions over and around such a location, and will publish the extent of the restriction through the Notices to Airmen (NOTAM) information system. When this situation exists, the rule is simple: Remain well clear unless you have official business there, and obtain permission from the controlling authority if your presence is required.

Air traffic controllers, the guys and gals in the towers, centers, and radar rooms all across the country, have been erroneously tagged "the traffic cops of the air." Marshals, maybe, but not policemen. They issue instructions and clearances to maintain a safe, orderly flow of air traffic. System success (that's where everybody gets his flying done and nobody gets hurt) depends on timely, reasonable instructions from the controllers, and prompt compliance on the part of the pilots. When, for example, a controller asks you to enter a left downwind for a particular runway, don't take it upon yourself to disregard his request and enter a *right* downwind. Such unilateral actions are not only dangerous, they're specifically prohibited by the rules. Of course, there's always that blanket rule about pilot responsibility, but it's reserved for emergency action, and you are expected to let the controllers know right away when in your judgment you must deviate from an instruction or clearance in the interest of safety.

Although the voice from the tower will almost always specify the traffic pattern in the landing information, the accepted standard is left-hand turns; this is also applicable to airports without control towers. On occasion, local conditions such as high terrain or obstacles on the left-hand pattern side of the runway, or a parallel runway in simultaneous use, will require right turns before landing or after takeoff; when in doubt, ask the controller. Nonstandard traffic patterns at uncontrolled fields are indicated by appropriate light signals or visual markings.

FLIGHT RULES FOR VISUAL CONDITIONS

The aviation community shortened Visual Flight Rules to VFR a long time ago; it's the pilots' way of saying the weather is good enough to proceed in accordance with those rules that allow visual separation of airplanes. The general operating rules obviously cover all of the things a pilot would want to do when he's able to see where he's going, so there is little remaining except those regulations that establish just what VFR really is.

All of the special situations notwithstanding—controlled versus uncontrolled airspace, inside or outside a control zone, helicopter operation, and so on—a pilot can't go wrong by always observing the longstanding basic rule of 1000 or 3. Whenever the ceiling (the lowest layer of solid clouds) is less than 1000 feet above the ground *or* you can't see at least 3 miles from where you stand at the airport, don't expect to have enough time to separate yourself from other airplanes. Not much to begin with, the effective visibility is reduced considerably when you're chasing the edge of a murky 3-mile circle at 100 miles per hour or more.

In uncontrolled airspace, you may legally operate with as little as 1 mile visibility so long as you can remain clear of all clouds—but, of course, you have absolutely no assurance that there's no one else coming your way. It's a calculated risk at best; you'll have to decide if it's justified.

The instrument pilot, wrapped in his security blanket of clouds, knows that another airplane will never pass closer than 1000 feet above or below—that's the responsibility of the air traffic controllers when pilots can't see. Under VFR conditions,

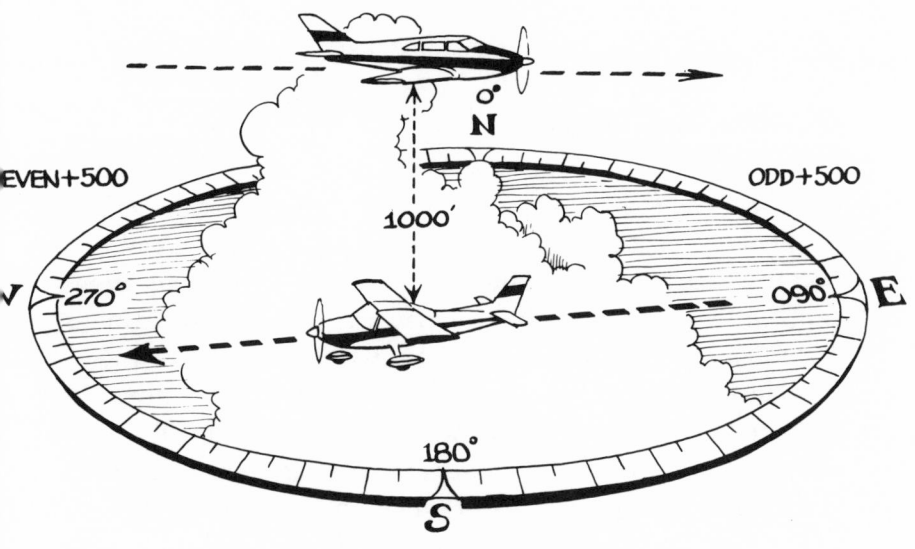

Figure 51

VFR vertical separation

pilots are required to provide their own vertical separation
through adherence to a clever system of altitude assignments
depending on direction of flight. Whenever your magnetic
course lies between 0° and 179°, the rules call for any *odd* altitude plus 500 feet; for example, 3500 or 5500 or 7500. Fly the
other direction if you will (a magnetic course between 180° and
359°), but change your flight altitude to an *even* thousand plus
500 feet—as in 4500 or 6500 or 8500. If all pilots go along with
the rule, no one will ever have less than 1000 feet between VFR
airplanes. Of course, vertical separation depends on proper altitude information, so you must always adjust your altimeter to
the setting obtained from a station within 100 miles along your
route of flight.

Many of the visual flight rules, and particularly those dealing with operations at a controlled airport, depend on communications between pilot and tower—but what about the
unhappy situation when the radios fail? If it's only the receiver

that's gone belly up, tell the tower your problem and ask for light signals; they have a powerful spotlight with which a beam of appropriately colored light can be shot right into your cockpit. (Of course, you must know what these visual signals mean; see the next chapter, "All About Airports," for the code.)

Should the entire radio system blow its electronic mind and you can't let the tower know that you have a communications problem, you are officially excluded from that airport. Unless it's an emergency situation (or about to become one), you've no choice but to land at a non-tower airfield. If you really need to fly to the airport with the tower, a telephone call will almost always result in permission to enter the traffic area and land on the basis of light-gun signals. It's no problem for the controllers when they know you're coming; it's the surprises that give them fits.

So much for a quick tiptoe through the regulatory tulips. The scope of the flying rules is too wide, the changes and additions too frequent, for any pilot to consider himself current. It's a lot like the fellow who had a full-time, one-man job painting a bridge—by the time he had worked his way across the span once, it was time to start again on the other side. In the case of the bridge painter, it's job security; for the pilot, it could mean survival.

All About Airports

12

A landing place can be just about anything a pilot wants it to be, from a grass strip that serves his personal flying needs to a giant public facility certified by the federal government and administered as the big business it really is. There are more than 12,000 officially recognized airports in the United States, plus who knows how many landing sites used exclusively by their owners.

There's no doubt that the best learning situation is at a small, quiet airport where a beginner can give nearly all his attention to handling the airplane, but when the fundamentals are mastered, a pilot must progress to a larger airport. Today's pilot-certification requirements include operations at a busy terminal, whether or not the flyer intends to use such facilities. Size generally implies a more sophisticated airport, more people to talk to on the radio, more traffic to watch for, and, in general, a more complex piloting task, but all airports have certain commonalities, an understanding of which should help ease the consternation and confusion when you fly to Metropolis International.

The most obvious feature of any airport is its runways, or the lack thereof. When local interests wish to attract air commerce or modify existing facilities to accommodate larger, faster aircraft, the runways grow in length, width, and number. "Overbuilding" is usually the name of the game, since concern for safety dictates considerably more runway than is actually required for taking off and landing. As a matter of fact, airline operators may plan to use only 60 percent of whatever runway is available on landing—so a jumbo jet that requires 6000 feet to land and come to a stop needs a runway at least 2 miles long.

While you can find runways pointed in every direction on the compass, the predominant alignment in the United States is east-west, because of the prevailing westerly winds in our latitudes. Runways are universally identified by magnetic direction, the numbers reduced to the nearest ten degrees. Therefore, a runway that lies precisely on a north-south line with relation to the magnetic pole would be known as "Runway 36" for landings and takeoffs to the north, and "Runway 18" for a southbound pilot. The last zero is always omitted, and the first one is omitted more often than not, so "Runway 4" could actually be pointed anywhere from 36 to 45 degrees. If you fly to an airport you haven't visited for years and find that the old familiar "Runway 27" is now "Runway 28," they haven't rotated the airport. Instead, the magnetic variation has changed enough in that area to put the runway alignment closer to 280 degrees than to 270, and it's easier to renumber the pavement than to fight Mother Nature.

Taxiways and access roads are simply "there" on most airports, but on the larger fields the proliferation of concrete and asphalt requires detailed identification, usually by alphabetical designators that start arbitrarily from one end of the airport. At the other extreme, many very small airports have no taxiways at all; you'll have to use the grass along the side, or the runway itself. It goes without saying that such an operation must be conducted with the utmost regard for other aircraft in the process of taking off or landing—one at a time, folks.

Directional indicators—the numbers—are academic for most flight operations. If you're approaching a single-runway airport from the east and looking down the center line of the runway, the runway must be within a few degrees of 270 to be called "Runway 27." In the case of parallel runways, the one on your left will be identified with the letter "L" and that on the right with "R"—as in "27L" and "27R." Approach the runways from the other end and you'll find "09R" and "09L" in the appropriate places.

Runway markings become an integral part of an instrument approach procedure, for easier orientation during that critical period of time between sighting the landing surface and actually touching down. You'll find that runways that serve as the terminus for a non-precision approach procedure (such as VOR, Localizer, NDB) will have the landing threshold well indicated so that instrument pilots won't undershoot. When a precision approach system is installed (one with a glide slope), the runway needs to be more clearly identified to accommodate the lower minimum altitudes of such a procedure, so side stripes are then added, plus touchdown-zone markings, a 1000-foot marker, and an array of lights just prior to the threshold to aid pilots in visual orientation just before landing. These runway markings are usually done in reflective paint to make them more conspicuous at night and under conditions of very low visibility.

The boundary between taxiway and runway is clearly marked with "hold" lines, unmistakable double yellow stripes across the taxiway just short of the runway. You must hold on the taxiway side of these lines until the tower controller clears you onto the runway, and you are expected to move past the hold lines to be considered clear of the runway after a landing.

Parking lines have no official status with regard to providing a clear area in which to move an airplane, but they are intended to provide orderly parking and access for the types of aircraft that use the ramp. However, be aware that guidelines painted between rows of parked airplanes don't guarantee wingtip clearance, especially if you happen to be driving a machine with longer wings than most. When in doubt, have someone check the clearance for you, or shut down the engine and see for yourself.

FIELD ELEVATION

Always expressed in feet above mean sea level, the published airport elevation is the result of a surveyor's efforts at one location—the Airport Reference Point. The fact that you may be parked some distance from this bench mark could account for a difference in published elevation and what your altimeter reads when it is properly adjusted; few airports are absolutely level. Still, use of the published figure will keep you well within the bounds of accuracy for traffic patterns and altimeter checks.

AIRPORT LIGHTING

Somewhere out there, buried in the lights of the city, is an airport; the familiar open-field, grass-and-concrete pattern that is so easy to find during the day becomes just another unlighted expanse when the sun does down. The first thing to look for is the rotating beacon, which is usually mounted as high as possible on a hangar, the tower, or sometimes a nearby hill. The sweeping beams of white and green light that turn into flashes when viewed from the side indicate that the airport has at least one lighted runway. (Thought you saw two flashes of white light between each green? That's no illusion—that's a military airfield, and Uncle Sam would rather you went somewhere else to land.)

When there were more floatplanes around, lighted water landing areas were heralded by white and yellow rotating beacons, but don't expect to come across many of those today. The white-green-amber beacon that identifies a lighted heliport is becoming more popular, however, as the rotary-wing crowd ventures out more and more after dark.

Regardless of the colors, rotating beacons are hard to find unless the airport is out in the boondocks, away from the visual confusion of suburban sprawl. Once you realize that it's the flashing of the light that catches your eye, you'll make it a habit to look in the general direction of the airport and wait for one of these pinpoints of light to become apparent.

The rotating beacon serves a useful purpose during daylight hours as well—when it's operating between sunrise and sunset, the field is IFR (or someone forgot to turn off the light).

Runways are always outlined with parallel rows of white lights, and when there's a control tower on the field, the intensity can be adjusted by the controllers—maximum brightness for penetrating low-visibility conditions, lower settings to make it easy on the pilot's eyes down close to the runway. Certain visual illusions will be heightened in the presence of very bright lights, and you should never hesitate to request a lower setting when you can't see the runway for the lights.

Maybe the Great Electronic Garage Door Opener Revolution had something to do with it—at any rate, an increasing number of airports have installed controls that enable the pilot to activate runway lights by keying his microphone a certain number of times. There are varying combinations of lights available (usually just the runway lighting, although some systems include approach lights), nearly all of them operating on frequency 122.8, and they are designed so that *something* will turn on when you key your mike 5 times in 5 seconds. At

---

some installations, once the device is activated it's possible to increase or decrease the intensity by further coded keying of the microphone; detailed instructions are published with the airport information. All pilot-controlled lighting systems are designed to remain lighted for 15 minutes after each activation, and you should expect to be able to light up the landing area from approximately 15 miles away.

There's a whopping difference in the electric company's bills to a single-runway, VFR-only airport and a big-city, all-weather terminal. On the one hand, just a few white lights to mark the landing area; on the other, a complex and extensive arrangement of lights to aid pilots in their visual orientation when they break out of the murk as low as 100 feet above the runway. A fully equipped facility may have flashing strobe lights that identify the end of the runway, a spread-out pattern of approach lights to provide roll information and heighten depth perception, a series of sequenced strobe lights that take on the appearance of a ball of fire shooting toward the runway, and an array of lights buried in the touchdown zone of the runway itself, lights designed to continue giving pilots information during the touchdown and rollout.

Throughout the airport system, you'll find taxiways pinned to the ground with rows of blue lights (most multiple-runway fields provide lighted signs to direct you to ramps and runways), and all runway thresholds are indicated by green lights.

Of inestimable value to the night flyer, Visual Approach Slope Indicators (VASI) project one or more beams of color-coded light into the final approach area for vertical guidance during descent to the runway surface. VASI systems were originally intended to provide last-minute guidance to the pilot transitioning from all-instrument to all-visual clues, but they have since become accepted as safe, positive aids to all pilots. There's great peace of mind in knowing when you're on a glide path that will carry you safely over all obstacles and terrain on the way to the runway. Such guidance also wipes out the too-long, too-fast landings likely to result from reliance on raw visual inputs at night.

There are three types of VASI systems in general use. The simplest is a tricolored light beam from a single projector mounted at the left side of the runway near the approach end. On final, you'll see a green light when you're on the proper glide slope, a red light when too low, and amber to let you know you're higher than you should be.

A two-light VASI requires pilot interpretation of the red-white color combinations he sees from two projectors installed just to the left of the runway, again near the approach end, one slightly behind the other. When two white lights are perceived, the airplane is too high; two red lights indicate a position below the proper glide path, and the combination of red from the far light and white from the near light means that the airplane is precisely where it should be. With either system, set up an attitude to produce the final approach airspeed you want, and "fly the lights" with power. Too high? Reduce throttle setting a bit and watch for a change in color. When red lights signal danger, add some power and fly back up to the glide path.

Runways that serve jumbo jets will frequently be equipped with a three-light VASI system, to accommodate the farther-down-the-runway touchdown point of these huge airplanes. On final, you'll see three lights beside the runway, projecting the same red-white combinations, which result in *two* visual glide paths—the upper one for the jumbos, the lower one for lighter aircraft. Your task is to determine which glide path to use and how to interpret the appropriate pair of lights. *Caution:* if you're following a jumbo to the runway, always use the *upper* glide path; the wake turbulence from the larger airplane is found below its flight path, which you can bet will be exactly on the upper VASI beam.

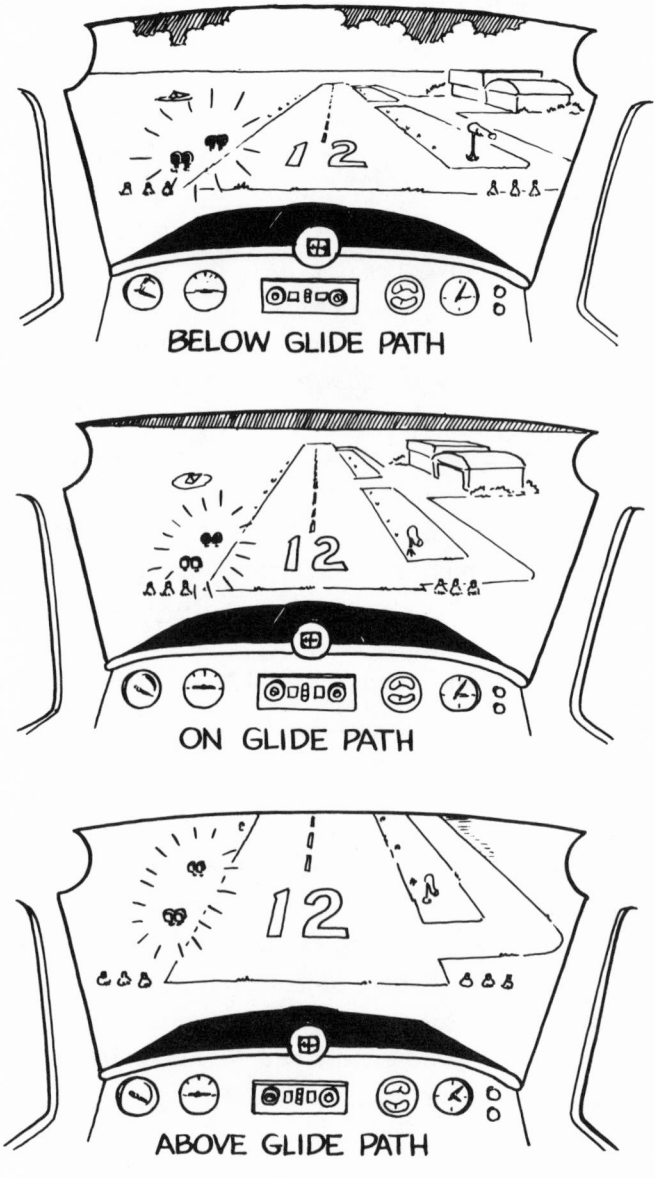

Figure 52

Pilot's-eye view of the VASI system at work

The entire concept of air traffic control in visual conditions is based on the pilot's being provided with information that will help him see and avoid other aircraft. While en route, or when flying to or from an airport with no radio facilities at all, it's strictly an eyeball operation . . . and faith that everybody else is following the rules of the road. Here's where it really pays to fly at the proper altitudes, execute left-hand turns in traffic patterns, and do all the other standard maneuvers that your fellow pilots expect of you.

At the first level of radio-aided traffic control is the airport with a unicom facility, which is intended not for official traffic control purposes but for advisories between pilots and the folks on the ground. Inbound to such an airport, it's good practice to announce your position while still fifteen miles or so away, and you'll usually get a reply that lets you know the runway in use and who else is flying around the airport. Treat those responses with caution, however, because anyone who happens by could pick up the mike and give you the numbers. If you don't get a reply right away, don't keep calling—there may be only one person on the field, and he's busy fixing an airplane or pumping gas. There's no substitute for your own vision and good judgment.

Pilots should use unicom (122.8 MHz at non-tower airports) in the traffic pattern on the basis of common sense and the realization that too many transmissions become counterproductive. At the normal traffic pattern altitude of 800 feet, your unicom transmissions are being heard 35 miles away over flat terrain; add up all the airplanes flying around little airports on a sunny Sunday afternoon, and the interference makes it tough for *anybody* to get a word in edgewise. A better procedure is to announce your position and intentions, then keep your eyes open and your mouth shut.

The second level of air traffic control is provided by a simple control tower, frequently a part-time operation. With a tower, an Airport Traffic Area is established, within which pilots are furnished more information relative to airport conditions and other traffic using the field. Such a tower usually has a *local controller*, who keeps tabs on all the airplanes in the Traffic Area and who issues clearances for takeoff and landing. While the local controller advises the pilots, it's still the people at the flight controls who are responsible for staying away from other airplanes. A *ground controller*, communicating on a different frequency, oversees taxiing operations on the airport surface.

To start a typical flight from a "second-level" airport, contact the ground controller and request taxi clearance. He'll tell you which runway is in use, advise you of the wind and altimeter setting, and clear you to that runway. (If you know what the runway, wind, and altimeter setting are from listening to his transmission to another pilot, let the controller know; that saves time and radio chatter. The standard phrase is "I have the numbers.")

When your before-takeoff checklist is completed, switch to the tower frequency and let the local controller know you are ready. The response will usually be one of three: "Taxi up to and hold short of the runway" (that means the double yellow lines that separate runway from taxiway); "Taxi into position and hold" (move onto the runway and get lined up but don't start your takeoff yet); or "Hold your position" (there's probably someone on final approach). When the controller sees that a proper clearance between airplanes on the runway can be maintained, you'll be "Cleared for takeoff."

Once your plane is off the ground, controllers assume it will make a standard departure (if you're not sure what's standard for the airport, *ask*), and nothing more needs to be said. When you're beyond 5 miles, you can turn off your radio if you wish.

If you are approaching an airport with minimum air traffic

control facilities, good practice calls for a report 15 miles out. Tell the controller who you are (your type of aircraft helps him sort out his traffic), where you are (use exact locations wherever possible), and what you'd like to do. "Downtown Tower, Cherokee so-and-so at the grain elevators, landing"— make it just that simple. Tower will come back with the numbers (if you have them, let him know on the first call) and instructions for entering the traffic pattern. Sooner or later, you'll be cleared to land, and you're expected to turn off the runway at the first available taxiway. Once past the double lines, you should contact ground control for clearance to the parking area. If you're unfamiliar with the airport, don't hesitate to ask the controller for help—that's one of the things for which he's paid.

At the third level of air traffic control complexity, business has increased to the point where repeating the numbers to each and every pilot takes up too much time on the air. Such a tower makes the information available through a continuous recorded broadcast called Automatic Terminal Information Service—ATIS, known to pilots as "ay-tiss." When you select the appropriate frequency, which is published on the navigation charts and wherever basic airport information is found, you will hear something like: "This is Executive Airport, information Golf. Weather clear. Winds two eight zero at seven; altimeter setting three zero three one. Runway two five in use. Advise the controller on initial contact you have Golf . . . *bleep*." (Each time there's a change in the weather or the runway or anything that would be significant to pilots, a new recording is made and the alphabetical identifier—in this example, "Golf"—moves up one letter. The controller doesn't say *bleep*, but between each repetition of the recording you'll hear a tone that indicates that it's about to repeat itself.)

Listen well in advance when ATIS is working, and include your knowledge of the numbers in your first transmission:

"Bonanza so-and-so ready to taxi from the main ramp with Bravo," or "Skymaster so-and-so ten west with Charlie, landing." Now the controller knows that you know what's going on, so your instructions or clearance will likely be very simple and direct; "Bonanza so-and-so cleared to taxi," or "Skymaster so-and-so, report on left downwind." ATIS saves a lot of time and chatter.

The fourth step in airport sophistication is the addition of radar service to all the facilities mentioned beforehand. The ATIS broadcast usually includes a radio frequency on which to obtain radar service when approaching or departing the terminal area. In addition to telling you the whereabouts of other radar-identified aircraft that might get in your way, the radar controller will provide vectors (headings to fly) and altitudes to get you into the immediate vicinity of the airport, where you'll be turned over to the tower for a landing clearance.

Departure from a radar-equipped airport usually involves contacting *clearance delivery* before taxiing (the frequency is probably included in the ATIS broadcast) to obtain a heading and altitude to be flown after takeoff. At some predetermined point, the tower controller will hand you off to a departure controller, who will use his radar equipment to help guide you safely out of the area.

Terminal radar service is great—so long as all its users realize that it's no substitute for a good set of eyeballs in every cockpit. Since this service is voluntary, there is always the very good possibility of some aircraft in the terminal area completely unknown to the controllers. Those are the ones *you* have to look out for.

At the top of the airport heap are those terminals that are so busy that a Terminal Control Area—the upside-down wedding cake—has been imposed around them. The most significant restriction about them is the mandate that all inbound and outbound flights obtain a clearance before entering or departing the TCA, in effect a "fourth-level" airport at which you have no choice in the matter of accepting radar service. The very busiest of these big-city airports—the Group I TCAs—accommodate such a steady stream of large, fast aircraft that student pilots are prohibited, and low-experience light-plane operators will usually feel much more comfortable using one of the smaller airfields that serve the city.

WHAT TO DO IF THE RADIOS QUIT

Common sense dictates that a pilot without two-way radio-communication capability do something other than land at a tower-controlled airport. Regulations make such landings illegal. Entering a landing pattern as "the phantom ship" does nothing but make the controller's job more difficult, as he must guess what you're going to do next. Far better would be a landing at an uncontrolled field close by; a telephone call to the tower will almost always result in permission to enter the airport traffic area sans radio, but then the controllers know who you are, where you're coming from, and what you intend to do.

When your no-radio intentions are known, or if the radios quit while you are flying in an airport traffic area, there's a system of visual communications available from every tower cab. The controller can aim a light-gun at your airplane and signal with a beam of red, white, or green light. Of course, the pilot must be looking at the tower if the system is to work; when you get the message, rock the wings or flash the landing lights in acknowledgment. Here's the universal code for tower light-gun signals:

Signal	On the Ground	In Flight
STEADY GREEN	Cleared for takeoff	Cleared to land
FLASHING GREEN	Cleared to taxi	Return for landing; to be followed by STEADY GREEN at the proper time
STEADY RED	Stop	Give way to other aircraft and continue circling
FLASHING RED	Taxi clear of runway	Airport unsafe—do not land
FLASHING WHITE	Return to starting point on airport	(Not used in flight)
ALTERNATING RED AND GREEN	General warning signal: exercise extreme caution	

TRAFFIC FLOW

The active runway at almost any airport is determined by the wind direction, since taking off and landing into whatever wind is available offers performance benefits. Pilots are informed of the runway in use by the tower or the ATIS broadcast at a controlled airport, by whoever answers on the unicom frequency at a field without a tower, and by personal observation when there's nobody else around.

For reasons of noise abatement, or more efficient handling of traffic by the controllers, some airports designate a calm-wind runway, to be used in preference to the others unless the wind direction/velocity combination makes it unwise. So it's possible to be cleared for a downwind approach and landing on a day when the air is moving slowly—not a bad deal at all, so long as you recognize the situation for what it is, and adjust your flight techniques accordingly. The VASI system is most helpful in such a set of circumstances. As a matter of fact, there's absolutely nothing wrong with *requesting* a downwind landing or takeoff on a sufficiently long runway when that would save you considerable taxi time either coming or going.

Where there's no one to let you know which way the wind is blowing—as is the case at many small airports—the management provides either a wind sock or a wind tetrahedron; the latter is conveniently shortened to "wind tee" in the parlance of aviators. The sock, nothing more than an open-ended fabric cone that streams out in the breeze, is a reliable source of information about wind direction and velocity.

The wind tee is an elongated three-dimensional triangle with enough aerodynamic properties to make it point into the wind, thus indicating the preferred direction for takeoff and landing. Sometimes the tee is modified to look like a small airplane mounted on a swivel close to the ground, which makes the interpretation even less difficult—use the runway most nearly aligned with the indicator. Both wind socks and wind tees are normally lighted after sunset.

When no formal air traffic control exists, all aircraft are expected to follow a general pattern of left-hand turns around the airport when approaching for landing. In some cases, the airport management has decided that right-hand circuits are better (perhaps to keep airplanes away from towers or hills, or maybe for community-relations purposes); such a nonstandard pattern will be indicated by a segmented circle somewhere on the airport surface. When in doubt, look for the pattern indicator and set yourself up accordingly; that's better than blindly entering a left-hand pattern every time and suddenly finding yourself nose-to-nose with another airplane going in the proper direction.

Traffic patterns exist for two very good reasons: first, to pro-

---

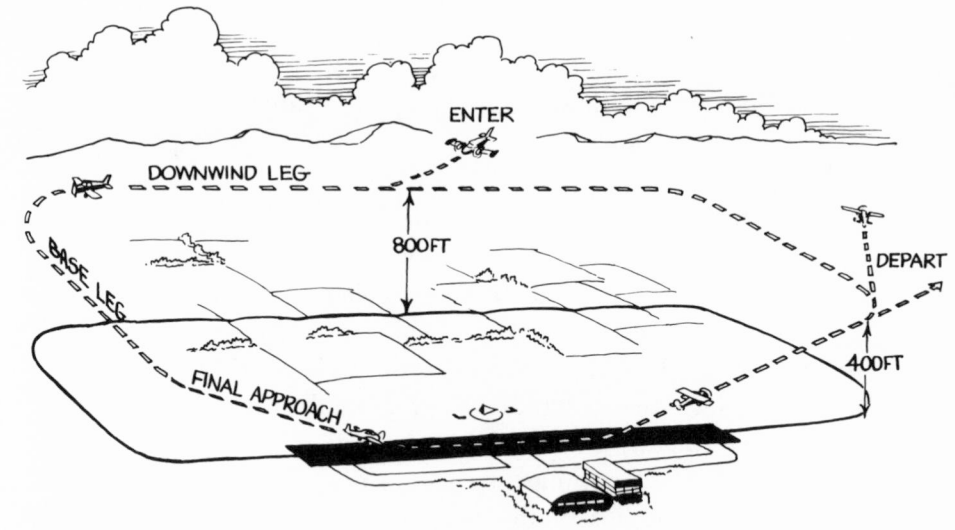

Figure 53

Standard traffic pattern

vide an orderly traffic flow when pilots remain in the immediate vicinity of the airport for the purpose of practicing takeoffs and landings; second, to provide a sensible and safe procedure for entering and leaving the airport area. So that all pilots will know what to expect, it is important that each aviator observe a certain amount of standardization in his pattern techniques. Described by an aircraft taking off with intent to remain in the pattern for practice landings, a standard pattern would consist of:

THE TAKEOFF LEG, a continuation of your course down the runway, climbing to approximately 400 feet above the ground, at which point you would execute a 90-degree left turn onto

THE CROSSWIND LEG, continuing the climb while flying perpendicular to the runway. When you are about a half mile away, from the airport, turn left 90 degrees onto

THE DOWNWIND LEG, which should be flown at 800 feet above the airport elevation (altitude on downwind is sometimes modified up or down to suit local conditions), parallel and opposite to the direction you intend to land. When roughly abeam the landing end of the runway, another 90-degree left turn puts you on

THE BASE LEG, for the purpose of flying over to the center line of the runway. You should be descending all the while, and when lined up with the runway, turn 90 degrees left once more and you're on

THE FINAL APPROACH LEG, which eventually leads you to the desired touchdown spot.

Standardization like this results in a predictable traffic flow, with each leg providing time and positioning for pilots to visually clear the airspace they are about to enter.

Traffic patterns are generally wider in a pilot's early stages of training, reflecting the increased time required to plan and execute all the things that need to be done. Faster aircraft can be expected to fly wider patterns, and it's not uncommon for the tower controller to request slight variations in pattern dimensions for the purpose of spacing—especially when the traffic mix includes airplanes of significantly different airspeeds or performance capabilities. Airliners, business jets, and other larger aircraft seldom fly traffic patterns in the light-plane sense—they are usually operating under instrument rules and proceed straight in to the airport.

Although there is no documented standard procedure for departing the traffic pattern, it makes good sense to get out of that potentially congested area as soon and as safely as possible. Fly a normal takeoff leg, make sure there's no one else ahead of you, then continue straight ahead or depart with a 45-degree turn. Either way, conflicts with inbound airplanes are avoided . . . *if* everyone follows the rules!

Though not standard, this technique should be considered a

normal airport departure, anticipated by controllers and other pilots—it's much safer than any other departure at uncontrolled airports. Where there's a tower, variations are quite acceptable if the controller agrees: depart straight out, turn right instead of left, whatever is most efficient for the circumstances at hand.

Adherence to an accepted procedure for entering a traffic pattern is every bit as important as the departure. By approaching the airport on a 45-degree entry leg aimed at the midpoint of the landing runway, incoming pilots give themselves time to search the airspace ahead for other airplanes, and they will be well clear of departing flights; the two procedures are complementary. Such an entry leg not only provides invaluable scan time, it is anticipated by other pilots in the pattern. When the downwind leg is intercepted, turn to join the normal flow of traffic and proceed as though you had been in the circuit all the time.

There are variations in the entry procedure at an uncontrolled airport. For example, a pilot approaching from the side opposite the downwind leg would be wise to overfly the airport well above pattern altitude, clear the entry area, and set up a 45-degree course to the downwind when he reaches the other side. An entry on the base leg is somewhat less desirable, since you will have less time to clear the downwind and final legs. A long straight-in approach is an invitation to disaster at an uncontrolled airport.

Add a control tower, and the pattern entries vary all over the lot—but now there's someone who knows where all the other airplanes are, and the controller will do his best to get you on the ground with the least possible delay. You will encounter frequent deviations from the familiar left-hand rule: right turns, base-leg entries, even straight-in approaches—whatever suits the situation best.

SAFETY IS AN
ACTIVE SET OF EYEBALLS

The traffic pattern has become the most popular spot for mid-air collisions—airplanes come together more frequently here than anywhere else in the entire airspace system. Whether you're approaching a deserted country grass strip at seven o'clock on Monday morning or departing Kennedy International at the height of the rush hour, don't trust the standard procedures or the controllers to the point of neglecting to look around. Just like the unloaded gun that somehow fires at the wrong time, the airplane that shouldn't be there can ruin your whole day.

Part Four

HUMAN
ELEMENTS

Pilot Certification and Currency
You Want to Fly a Tail-dragger?
Psychological Factors and Emergencies

Pilot Certification
and Currency

13

"This part prescribes the requirements for issuing pilot and flight instructor certificates and ratings, the conditions under which those certificates and ratings are necessary, and the privileges and limitations of those certificates and ratings." So reads the very first paragraph of Part 61 in the Federal Aviation Regulations, the undisputed bible of the airman's world when it comes to achieving, maintaining, and improving one's status as a pilot. Though a handful of the sovereign states extract a buck or two from their flying citizens' pockets for the privilege of becoming an aviator, anyone who pilots an aircraft of United States registry must first be properly certified under the federal laws.

Part 61 covers all facets of certification and currency, and therefore provides at least minimum-standards control over the quality of a pilot's initial training and the maintenance of his knowledge and skill. The regulations define very clearly those areas of aeronautical knowledge, skill, experience, and medical qualification required for the various levels of aircraft piloting. By setting higher standards for the advanced certificates and ratings, the laws tend to protect noncommercial flyers from themselves, and to justify travelers' confidence in the aptitude of the person in the left seat.

The frame on which the documentation of flying privileges is built is the *pilot certificate*. There are four levels: student, private, commercial, airline transport pilot, with flight instructors occupying a separate niche in the rulebooks. The student pilot certificate is simply a license to learn the basic flying skills, and as such has no additional ratings. Further pilot qualifications (at any level) must be described by one or more *aircraft ratings: airplanes, rotorcraft, gliders or lighter-than-air*. An *airplane* pilot may be qualified to operate single- or multi-engine aircraft on land or sea, or any of the possible combinations thereof. A *rotorcraft* rating is awarded with regard either to helicopters (powered rotors) or to gyroplanes, those nearly extinct machines whose rotary wings turn freely in the airstream while the thrust comes from a conventional propeller. There's only one kind of *glider* rating, but pilots are restricted to the type of launching method (ground or aero tow) in which they are officially trained. Airships (blimps) and free-flying balloons make up the *lighter-than-air* category.

The complexities of a "large" aircraft (one that weighs more than 12,500 pounds at takeoff) and any turbojet-powered airplane (one with pure jet engines—no propellers) require *type ratings*, earned through in-flight demonstrations of knowledge and competence in the operation of such an aircraft. The award of a type rating means that the official authorization to operate a specific type of aircraft (a Boeing 707, for example, or a DC3, or a Learjet) will be added to the pilot certificate.

Perhaps the most useful pilot credential is the *instrument rating*, added to either a private or a commercial pilot certificate after a satisfactory demonstration of operations in the IFR world. In addition to successfully taking a detailed written examination, an instrument-rating candidate must have accumulated at least 200 hours of flying time, including 40 hours during which no outside references were used, only the aircraft instruments. Some of this experience may be acquired in an aircraft simulator, and a generous amount of the required instrument time is set aside for practicing the techniques and procedures required for the checkride.

REQUIREMENTS, PRIVILEGES, AND LIMITATIONS

A *student pilot* certificate is available to almost anyone who is in reasonably good health and can read, speak, and understand the English language. Obtainable at age 16 (glider and balloon students can get theirs at 14), the student certificate combines three documentary functions on one piece of paper: the certificate itself, proof of medical qualification, and space for the flight instructor to endorse solo and cross-country flights. Medical examiners (doctors with special qualifications in aerospace medicine) issue student-pilot certificates in conjunction with a Class Three medical certificate; both expire 24 months after issue, and if a student has not turned into a pilot by then, it's back to the doctor for another checkup and a new license to learn.

A student pilot operates on a short leash, at the other end of which is his flight instructor. A "note from teacher" in the form of written endorsement must be placed on the student's certificate and in his logbook prior to solo flight. In addition, each solo cross-country flight must be checked in advance and approved in writing by his instructor. The regulatory leash is not long enough to permit carrying passengers, flying for hire, or even flying in furtherance of a business. A student-pilot certificate is strictly for learning.

The difference between a student pilot and a *private* pilot is at least 40 hours of flight training (half with an instructor, half solo), enough book learning to pass a written examination, and an in-flight demonstration of aeronautical ability (otherwise known as a checkride). The flight training will have explored virtually all the maneuvers and techniques required for visual operations in the nation's airspace, plus enough instrument training to allow the beginning pilot to extricate himself from an inadvertent low-visibility situation. Minimum age for certification as a private pilot is 17. (Glider and balloon pilots once again get a break—their minimum age is 16.)

A private pilot's medical qualifications must be renewed every 24 months, but the *pilot* certificate is valid unless revoked, and the newly minted aviator is very much on his own. Within the limits of weather conditions and a host of operational rules, he may fly where, when, and how he wants to. Passengers may be carried (they may pay for their share of the expenses, but not one penny more) and the airplane may be used in business so long as it's just another way to get from here to there. The major restriction on a private pilot's aerial activities is the absolute prohibition of flying for hire—being paid expressly for carrying people or things in an airplane.

The type of aircraft in which most people earn their wings is largely a function of economics, which dictates the almost universal use of light two- and four-seat training airplanes. However, for the person whose needs (or wants) are better served by a helicopter or a twin-engine airplane, glider, balloon, or airship, the certificate can be earned in any one of those groups of aircraft. A number of people who didn't even know which end of an airplane to take hold of started right off in helicopters and are doing just fine.

The next level—the working level—is *commercial pilot* certification, with a minimum age of 18, an experience requirement of 250 hours (that's for airplanes; certification in other types of aircraft requires somewhat less), and the privilege of accepting compensation for pilot services. More complicated airplanes come in for their share of attention during commercial training, with at least 10 hours of experience required in an airplane with retractable landing gear, wing flaps, and controllable-pitch propeller. The commercial candidate is being prepared for bigger and better equipment.

Another written exam is in the offing for pilots who want the word "commercial" on their certificate. This one is considerably more difficult than the private test, and focuses on the regulations pertaining to "for-hire" flights—since a commercial pilot is a potential guardian of the safety of paying customers, he is expected to be aware of the limitations of the business. Prior to the checkride, a commercial candidate must pass a thorough oral exam dealing with aircraft systems, performance, and emergency procedures in addition to the basic knowledge of the airspace and flight operations in general. The checkride will be more demanding in scope, and the pilot will be expected to handle the airplane with much greater proficiency and facility.

When passengers pay for air transportation, they rightfully expect more than a "third-class" guarantee of the captain's health, and this is accomplished by requiring commercial pilots to visit the aviation sawbones at least every 12 months for a Class Two medical examination. It's a bit more involved, and the increased frequency of these personal inspections augurs well for the reliability of the individual in charge of the airplane. (A second-class medical certificate automatically reverts to third-class limits—24 months—whenever the holder is operating under private-pilot rules.)

A commercial certificate without an instrument rating, though obtainable, is relatively sterile; the pilot is limited to 50-mile trips when carrying passengers for hire, and if the folks who bought tickets need to fly during the hours of darkness, they'll have to find an instrument-rated pilot. On the other hand, there are a lot of aviation occupations that require a commercial certificate but are not at all concerned with IFR or passenger-carrying operations: banner-towing, agricultural application, pipeline inspection, helicopter patrol . . . the list goes on and on, and the laws relieve these pilots of the economic burden of obtaining a rating they don't need.

A great many pilots jump through the commercial hoops for no other reason than to gain official recognition of a higher level of proficiency. Though they probably never intend to earn their daily bread as aviators, these pilots have a certificate to show the world (and the insurance underwriters, who often look with great favor—and reduced rates—upon such an accomplishment) that they have climbed one step higher on the proficiency ladder.

The requirements for the *airline transport pilot* certificate— the revered ATP—make up the thorniest list of eligibility brambles in the business: The applicant must be at least 23 years old, a high-school graduate or equivalent, have a first-class medical certificate good for only six months, and, for the first time in the entire gamut of pilot qualifications, the ATP-to-be must be "of good moral character" (the book neglects to point out just how that is determined). The 1500 air-hours required of an ATP is broken into rather large chunks of night, instrument, cross-country, and pilot-in-command time, which insures a broad range of aeronautical experience in the candidates.

The ATP written exam may well be the toughest of them all, with near-total memorization of the air-carrier regulations almost a prerequisite. Other pilot tests may be taken at any time the instructor feels the student is properly prepared, but the ATP rules specify complete fulfillment of the flight-experience and physical-examination requirements before the test may be administered. It's the only pilot test for which an instructor's written authorization is not required.

A thorough oral examination of the applicant's knowledge of his airplane precedes the checkride, which amounts to a very demanding instrument procedures evaluation; the ATP certificate automatically includes an instrument rating.

Originally designed to qualify the captains of public air-carrier aircraft (they are still the only pilots in our aviation system who *must* be so certificated), the eminence and exclusivity of the ATP certificate has eroded with time. Pilots with absolutely no hope of holding down the left seat of an airliner obtain an ATP because there's nothing else beyond the commercial certificate to indicate progression of skills and experience for either personal or insurance reasons. The very name has suffered degradation at the hands of the aviation community; there are a lot more references to the ATR—implying that it's just another rating—than to the correct, full-fledged certificate designation, ATP.

Not only should he be a reasonably good guider of aerial machines, a *flight instructor* must possess the ability to instill good practices and techniques in others. His certificate is a document unto itself, but valid only when accompanied by his commercial and medical certificates. He must be at least 18 years old, have an instrument-rated commercial or ATP certificate, and pass written and oral examinations that cover thoroughly the many considerations of the learning process. The flight instructor checkride (which must be repeated for each additional type of aircraft—helicopters, gliders, and so on—in which he intends to instruct) is to find out not so much how well he can handle the aircraft as how well he can pass on his knowledge to a student.

In recent years, the flight instructor has been given almost total responsibility for the students who come to him to learn, from endorsements of their ground and flight training to final recommendation for a checkride. To make certain that the training philosophies and standards of the Federal Aviation

Administration are carried out, a flight-instructor certificate must be revalidated every two years. It's the only pilot certificate that expires from lack of use.

CURRENCY REQUIREMENTS

Pilot-population studies have borne out what everyone has suspected for a long time—that flying skills deteriorate when they are not exercised frequently and regularly. As a result, a number of rules on the books require certificated pilots to maintain what is considered a bare minimum of aviation experience.

The blanket rule decrees that no one may act as pilot in command of any aircraft unless he has successfully completed a Biennial Flight Review within the preceding 24 months. The BFR may be administered by any flight instructor or an FAA inspector, and is a two-part exercise: an oral review of current rules and procedures, followed by a session in the air, during which the reviewer has an opportunity to evaluate skills and techniques in the operations for which the pilot is rated. Any successful pilot-proficiency checkride—one that results in upgrading to a higher certificate or adding a rating—is an automatic Biennial Flight Review.

Pilots must accomplish at least three takeoffs and landings within the 90 days preceding a passenger-carrying flight; three nighttime takeoffs and landings must be executed for currency if the proposed trip will take place during the hours of darkness.

An instrument-rated pilot has additional recency-of-experience requirements: To enter the IFR system (that means accepting an ATC clearance and proceeding under the umbrella of their protection, whether the weather is IFR or not), he must have flown at least six hours of instrument time in the previous six months and accomplished six instrument approach procedures in the same period of time. Should these requirements not be met for two consecutive six-month periods, the pilot must pass an instrument-competency check—another IFR checkride—before he may exercise the privileges of his rating.

Staying current also means keeping records, because each pilot must be able to prove compliance with the currency laws. A personal logbook documents the dates, times, and places of each flight, as well as the type of airplane and the type of pilot time involved, such as pilot-in-command, instruction received, solo, or instrument time. In the eyes of the lawmakers, since the pilot is responsible for the safe operation of his aircraft at all times, he deserves full credit in the form of loggable flight time. So the official definition grants you flight time from the moment the aircraft first moves under its own power for the purpose of flight until it comes to rest at the end of that flight. Your logbook should reflect the entire time from leaving the chocks until you're back in place on the ramp—so you may have more pilot hours than you thought.

You Want to Fly
a Tail-dragger?

14

Congratulations—and welcome to one of the smallest groups in aviation today, the company of pilots who fly airplanes with the little wheel in the back. That landing-gear arrangement was considered conventional at one time, but the overwhelming popularity of the nose wheel has relegated the tail-dragger to the bush pilots and a few hardy souls who really appreciate the fine points of flying.

Of course, the nose-wheel airplane is nothing new. Orville and Wilbur's first creations sat level on the ground, and it wasn't until large-diameter propellers were required to realize the power of early slow-turning engines that designers lowered the tail and raised the nose to provide ground clearance. The fact that the prop tips are well above long grass and chuckholes makes a tail-dragger the rough-field pilot's delight even today; combine that with light weight and high-lift wings, and you've a combination that can't be beat by anything but a helicopter for really short and lumpy landing strips out in the boondocks.

With all these good things going for them, why are tail-

---

draggers so rejected by the aviation community? There's a bit of the "old-fashioned" stigma—a lot of them are covered with fabric instead of shiny aluminum, and they'll likely as not have control sticks instead of "steering" wheels—but the principal weapon used in the character assassination of these little airplanes is "They're *hard to fly*."

Baloney.

There has never been an aircraft as responsive, positive, and pleasant to fly as the tail-dragger. To be sure, flying it well requires a touch of finesse not needed with modern designs, but the aerodynamic honesty of a tail-dragger develops an awareness of the forces and feelings of flight that a pilot can find nowhere else. Because of the low speeds at takeoff and touchdown, tail-dragger pilots must be willing to use the flight controls to whatever extent is required—of course, a control stick makes that a lot easier.

Instructors who have worked both sides of the street will usually agree that better pilots are turned out by tail-draggers, because they must be flown *all the time,* from chock to chock. The low wing-loading (not many pounds of aircraft weight per square foot of wing) means that tail draggers are only slightly removed from gliders, so the standard approach to landing is accomplished power-off—the very best way to teach a student the ways of the wind, and its effect on an airplane.

Contrary to popular opinion, tail-draggers are *not* skittish and hard to fly, but you've got to pay attention. It can't be all that tough—several generations of pilots have cut their eye-teeth on tail wheels.

WITH ITS NOSE POINTED TO THE SKY

A tail-dragger (an airplane with the little wheel in *back*) is ready to fly even while it's tied to the airport; no rotation necessary here, for the built-in angle of attack starts to produce lift when the slightest breeze flows over the wings. That's one accepted takeoff technique—merely add power until the airplane has accelerated to an airspeed that generates enough lift at that angle of attack, and presto, you're flying.

The attitude of a tail-dragger as it sits on the ramp is much more important in terms of landing, however. As soon as you climb aboard, take a mental picture of the cowling, the bottom of the windshield, whatever part of the airplane structure you can relate to the horizon. That picture is the one you want to see just before the wheels touch the runway at the end of each flight. In a tail-dragger, the normal landing is a full-stall, three-point landing—and the perfect attitude to make that happen is right there in front of you. By duplicating that pitch picture just before the airplane stalls, you're guaranteed a three-pointer every time.

Absolute certainty that the area into which you're about to fly is clear of other airplanes is the most important rule in self-preservation, and taxiing a tail-dragger is often a pilot's first experience in this regard. The elevated nose is bound to obscure some of your forward vision, and if you need to S-turn a bit to see where you're going and make sure there's no one else out there, you've formed a good habit.

A virgin pilot, or one who is at the controls of a tail-wheel airplane for the first time, will find out early on that there's more to taxiing a tail-dragger than simply steering. In a tricycle-gear arrangement, the airplane's center of gravity is ahead of the main wheels, and any force that tries to turn the airplane will be countered by a tendency to straighten things out again. Not so with the tail-dragger: The center of gravity is necessarily *behind* the main wheels, and when the airplane starts to turn the CG tries to continue straight ahead. It's easy to get a lot more turn than you bargained for.

When you come to that first corner in the taxiway, apply just enough pedal pressure to get things started in the proper direction and let the airplane's momentum do most of the work. Remember that you must sooner or later overcome that tendency to keep turning, so begin stopping the turn well before the nose is pointed in the direction you wanted. Don't be discouraged, because everybody makes gross over- and undercorrections while they're learning; even though your taxi tracks may look like a snake crossing a mill pond, the touch will develop with time and practice.

Without a doubt, there's nothing more troublesome than the wind when you're taxiing a tail-dragger. Talk about weathervanes—you're astride one! Every breeze invites a tail-dragger to turn until its nose is pointed directly into the wind—an invitation the airplane will always try to accept. Use whatever rudder you need to keep the nose pointed where you want it to go.

There's no problem taxiing *into* the wind—keep the airplane straight and level with the same control movements you'd use if you were flying—and when you're moving downwind a simple rule will keep you out of trouble: *Dive with the wind.* For example, taxiing with the wind coming from behind and to your left, move the stick all the way forward and full right; the elevators will be down to keep the tail from lifting, and the wind's efforts to raise the left wing will be foiled by the lowered aileron on that side.

A strong quartering tailwind might require more rudder than your tail-dragger's got. When that happens, use the downwind brake as necessary. Since the wind will be very much to your advantage at takeoff time, don't fight the wind any longer than you have to. Ask for an intersection takeoff—use your resources wisely.

And keep in mind that there will be days when the winds are too strong or gusty to operate *any* airplane safely. Because of their high angle of attack on the ground, tail-draggers are particularly susceptible to high winds. They've even been known to take off and fly by themselves . . . for a very short time.

Accomplish the before-takeoff checklist just as you would with any other airplane, with the exception of holding the stick or wheel all the way back during the high-power portion of the engine runup; the combination of prop blast and wind will often be enough to raise the tail off the ground. And if your tail-dragger's seats are tandem (one behind the other) always be sure to make a very thorough, full travel check of the flight controls before takeoff, to be sure that nothing in the rear seat will interfere. It's a good idea to lash everything down in the back seat when you go out for a solo flight.

OKAY, AIRPLANE, *I'M* IN CHARGE HERE

There's no substitute for a good, positive attitude at this point. A tail-dragger has a mind of its own on takeoff, and the first few efforts may lead you to believe that there's a living spirit in the airplane bent on making a complete ass of any pilot who defies it. Not so. The problem encountered most often is under-controlling followed by gross overcontrolling. Somewhere in between, particularly with respect to rudder control, you'll find happiness and eventual mastery of the machine.

As soon as you add power, asymmetric thrust becomes very evident, and you can count on needing a footful of right rudder. The more powerful the engine, the more the airplane will try to veer to the left. The P-51 (North American Aviation's Mustang fighter plane of World War II fame) had so much power and so little rudder that it was impossible to add full power all at once and keep the airplane on the runway. The strong left-turn tendency of the tail-dragger is increased by gyroscopic forces as the tail comes off the ground, and, as likely as not, there will be a crosswind from the left on your first attempt, so if you need full right rudder, push it all the way to the stop—that's why the rudder pedals are there.

Reluctance to use all the rudder is common among beginners (that's the undercontrolling part of the problem), and the next thing that happens is a grim determination to push the pedals through the firewall if necessary to keep the nose pointed in the right direction—that's the overcontrolling part. An admirable ambition, but as the airplane gains speed down the runway, those full deflections of the rudder will get more and more effective, and you'll likely find yourself getting behinder and behinder. After a few tire-squealing near misses with the runway lights, you'll discover that the secret is applying just whatever rudder *pressure* is needed to keep the nose from swinging—just pressure, not pedal movement. Later, when you've developed a smooth, steady takeoff technique, you can chuckle while watching another neophyte fanning the air with the rudder; it will look as though he's trying to make the airplane go faster by flapping the tail, but be tolerant—you were there once too.

In your spare time during the takeoff roll, take another picture of the three-point pitch attitude with the camera of your mind; it's what you want to see just before touchdown on the other end of the flight.

The stick or wheel should be allowed to "fly" all by itself during the initial part of the takeoff, and when the airspeed gets high enough, the tail will start to lift off the runway of its own accord. *Let it*, until the nose drops a little bit (How much? Depends on the airplane, but six inches is a good place to start), then hold it right there. This procedure is a compromise between acceleration (there's a little less drag with the tail elevated), wear and tear on the tailwheel (that little tire really turns up the revolutions if you keep it on the ground), and a safety cushion during the transition from wheels to wings. If you maintain back pressure on the stick to hold the tail down, the resulting high angle of attack will get you off the ground a bit prematurely. This is not all bad—the technique is recommended for certain aircraft in order to maximize their takeoff performance—but the entire operation will be smoother, showing a little more class, if you allow the tail to rise somewhat. (To cement the picture of this slightly tail-high attitude in your mind, sit in the pilot's seat and have someone go around back and lift the tail six inches off the ground. That's what it should look like on takeoff.) Hold that attitude and the airplane will lift off when it's ready, probably so smoothly that your passengers will have to look to see if they've left the ground.

Once away from the ground, a tail-dragger flies like any other light airplane, but don't expect to make coordinated turns with your feet on the floor. Most of these old-timers were designed when pilots couldn't even *spell* "Frise," let alone know that it was a new type of aileron that all but eliminated adverse yaw. Here's the essence of flight again, the need for you to do more than one thing at a time. You must consciously coordinate the angle of bank and the rate of turn, which in most tail-draggers means a lot of rudder pressure—especially during low-speed climbs, stalls, and other practice maneuvers. Bothersome? Perhaps, but once you get the feel of coordinated flight, you'll expect it, demand it, make it happen in everything else you do with any airplane. It's good, solid training.

APPROACH AND LANDING

The end of a flight contains perhaps the finest expression of the tail-dragger's qualities as a basic trainer, because the low wing-loading (and usual lack of flaps) favors a power-off approach and full-stall landing. There's no better way to develop your skill for that—one hopes—once-in-a-lifetime day when the engine quits and you have no choice but a glide to the ground . . .

how comforting to *know* that you can do it. When you've learned in a tail-dragger, a forced landing is just another normal procedure with the choice of fields somewhat limited. And by getting good at full-stall, three-point touchdowns, you'll always be putting the airplane on the ground at the slowest possible speed—a valuable contribution to your survival in a rough-field landing.

A cut-and-try pattern on a no-wind day is the place to start. From 800 feet above the ground and a half mile—no more—from the runway, close the throttle and set up a glide. For starters, begin the turn to base leg when the runway numbers appear 45 degrees behind you; be sure to notice the point on the ground over which this turn was made—it will be a good reference for the next time around. Turn final when you're lined up with the runway, and see what happens, altitude-wise. Too short? Add a little power (emphasis on *little*—remember the low wingloading; a couple of hundred rpm's go a long way in a light airplane) until you are back on the glide path, and resume your descent to the runway. Should you turn final much higher than you'd like to be, continue the power-off glide and land long if it's safe to do so, and adjust the pattern the next time.

In that happy situation where you do everything right the first time around, remember where you reduced power, where you turned base and final. Expect it to happen just about that way on your second try. If you remember where you reduced power to start the glide and where you turned base and final, and if you maintain the same airspeed and rate of turn throughout, you've established a no-wind groove for your personal power-off approach and landing. Above all, remember what it *looked* like.

The only adjustment you need to make on a windy day is to alter your flight path to get to the same spot on the runway every time. With a wind blowing, the pilot who insists on reducing power at the same geographic point, who religiously turns base over the same farmhouse, who diligently maintains the same rate of turn and airspeed as he did when the wind wasn't blowing, will find himself woefully short of the runway. The solutions are two: Either carry some power to make up for the air miles lost to the wind, or shorten the entire pattern to get the same effect. It's your choice, but remember that every approach aided by engine power shaves a little bit off the fine edge of the emergency skills you've developed.

Besides, there's something very esthetic about that little airplane whispering slowly toward the ground, with the engine and propeller noises all but forgotten; it's like flying a real glider.

Landing a tail-dragger involves not a whit more skill than is required for any other airplane, but it does demand the pilot's attention at all times. As with a nose-wheel airplane, the objective is to effect a smooth transition from a powerless glide to a power-off stall just above the runway surface.

At the "artful" point, which must be developed through experience and several excursions into those areas that aren't artful, begin to assume the landing attitude. Recall exactly what things looked like when you first climbed into the tail-dragger, as you were taxiing, and when you started down the runway; you must now make the airplane assume that very same attitude. In a tail-dragger, this is not an artificial attitude, or one that has to be practiced; it's there every time you sit in the pilot's seat.

Again because of the low wing-loading, you'll notice that a little back pressure goes a long way in establishing the landing attitude. Control *pressure* is the key, not actual movement. Once the attitude picture looks right, make it stay—hold it, hold it, hold it until the airplane stalls and settles onto the runway in a beautiful three-point landing. It isn't all that difficult to do. The big hurdle is finding out exactly when to begin the roundout and how rapidly to accomplish it.

Make it an unbreakable habit to touch down with the stick or wheel all the way back to the stops and *keep it there*. At this point, you need all the help you can get for directional control, and firm pressure keeps the tail wheel glued to the ground for maximum steering effectiveness. Since the center of gravity is behind the directional pivot point (the main wheels), any turning will tend to increase as the airplane tries to swap ends; careful but authoritative use of the rudder is once again indicated. If you have to fan the air for a while to find just how much you need, do it.

Brakes are almost never needed in a tail-dragger if you adopt the full-stall landing technique. Put a little wind into the picture, and you could get out and run alongside at touchdown. On those rare occasions when the brakes become necessary, use them with caution; a tail-dragger is an inveterate end-swapper.

Crosswinds and tail wheels are not entirely incompatible, but the light weight and low speeds of these airplanes require frequent use of full control travel. There will be times when even a full throw of rudder and aileron won't be enough to counteract the drifting effect of a strong crosswind, and those are the days when lightweight airplanes in general should remain tied to the airport. But there's an avenue of escape, and it's known as a *wheel landing*.

The *pièce de résistance* of tail-dragger flying, the wheel landing is a demonstration of pure skill—luck doesn't count— and the pilot who makes consistently good wheel landings has gotten there by means of diligent practice and at least some measure of native aeronautical skill and finesse. You'll use wheel landings in strong, gusty winds, because the higher airspeed required makes the controls more effective; you'll use them when landing a heavy airplane, because the higher airspeed required gives you a bit more cushion for a smoother landing; and you'll occasionally use wheel landings just to

---

make yourself look and feel good—in other words, a tail-dragger pilot likes to show off once in a while.

A properly executed wheel landing will result in the main tires greasing onto the runway while the airspeed is still well above the stall; you will literally fly the airplane onto the ground. The extra airspeed, normally dissipated in the process of stalling in the three-point attitude, is retained a while longer for added controllability.

Set up the approach as before, but realize that this one is going to be slightly flatter in deference to the power setting you'll maintain throughout the approach and landing. Most pilots prefer a slightly higher than normal airspeed for the entire procedure—maybe ten miles per hour faster—but in any event, you'll be maintaining airspeed with pitch control and rate of descent with the throttle.

Close to the runway (What is close? You'll discover your own definition after a few tries), maintain the pitch attitude that produces the airspeed you want, and reduce the power very slightly, allowing the airplane to settle with no change in airspeed. Pitch control is of the utmost importance here: Let the nose come up and the tail-dragger will start to climb because of the increased angle of attack; let the nose drop and the airplane will settle faster—with this kind of closure rate, the main wheels will bounce off the pavement instead of gently rolling onto the runway. You might try pressing downward on the control stick to prevent large inputs from your arm muscles at this point.

Assuming that the rate of settling is under control, hold that attitude until the wheels touch, then apply a slight amount of forward pressure—just a touch—on the stick to soak up the inevitable reaction of the touchdown, and smoothly reduce the power setting to idle. There'll be some fancy footwork required for a few seconds, as the power is reduced and the tail settles, but use whatever rudder you need to keep the nose pointed straight down the runway. Until the tail comes down, you're riding along on the pivot point, and all tendencies to turn will be magnified. You've *got* to pay attention.

The wheel landing can be used to develop a fine touch with your tail-dragger, but of course there's a price to be paid: The additional airspeed that is part and parcel of this maneuver will increase the landing roll and keep you on your toes maintaining directional control.

MAXIMUM PERFORMANCE

By the time you get good at normal takeoffs and landings in a tail-dragger, you are just about as close to maximum performance as you'll ever need to be. This is especially true of takeoffs, for there's little to be done to make the airplane fly sooner; the takeoff roll can be shortened slightly by holding the tail on the ground until the airplane lifts off, but be aware that ground effect may fool you into thinking the climb will continue at that very low airspeed.

The full-stall, power-off landing will produce a rollout that will fit onto nearly all runways worthy of the name. When even that distance must be shortened, the answer is a power approach in the landing attitude—perhaps the finest expression of the technique of controlling airspeed with the stick and rate of descent with the throttle. As you approach the landing site, gradually adjust the pitch attitude to produce an airspeed just above the stall, and let the airplane down with power. When over the desired touchdown point, be sure you're close to the ground before you reduce the throttle setting—power is all that's holding you up, and when it's gone, plunk . . . you'll have landed right there.

DOWN WITH NOSE WHEELS?

Hardly, because there are a lot of things that a tail-dragger can't do. But, given the benefits of skill development, awareness of the effects of wind on an airplane, and the fact that you have no choice but to *fly* a tail-dragger every moment, it seems that most pilots would be better pilots after some training in an airplane with the little wheel in the back.

Psychological Factors
and Emergencies

15

In 1783, the king's word was law, and when Louis XVI of France requested a demonstration of the Montgolfier brothers' newly invented balloon, Jacques and Joseph rushed right over to Paris with a hot-air contraption big enough to carry a man, figuring to impress His Majesty and become the world's first flyers in one fell swoop. But when Louis realized that one of the brothers was actually planning to ride the balloon into the sky, he turned thumbs down and suggested that they substitute a trio of barnyard animals for the first flight. Louis's concern for the welfare of his loyal subjects may have been the first official recognition of the hazards of the upper air.

Early predictions of the physiological problems that surely awaited any man who ventured into the sky were founded in ignorance and steeped in superstition, but King Louis and others of his day were on the right track. There *are* some unexpected, unpleasant, and downright dangerous situations lying in wait for the unprepared aviator. After all, the human body has adapted itself to the conditions at the very bottom of an ocean of air, which means that each time you leave the ocean floor, you're asking the body and mind to perform at higher-than-normal levels of proficiency (flying an airplane requires considerably more skill than walking down the street or driving a car) in a suddenly and sometimes drastically changed environment.

HYPOXIA

You've become accustomed to a certain amount of air pressure pushing oxygen from the air into the bloodstream, which circulates the oxygen around the body to be used as needed. When increased activity demands more oxygen, the speed of the pump is automatically turned up a bit, and your breathing rate increases—again automatically—to move more oxygen-laden air into the lungs, where the exchange takes place. Through delicate sensing of blood chemistry, your body will make these adjustments entirely on its own whenever the signal flags go up, which means that as you climb through the atmosphere (walk up a mountain, take a balloon ride, or fly an airplane), you undergo a certain amount of physiological change.

The body is able to accommodate to a decrease of pressure quite handily in the lower levels, but sooner or later a point is reached at which certain parts of the system begin to lose efficiency because of a lack of sufficient oxygen. This condition is called *hypoxia*. Sight will be among the first functions to go; a slight impairment of night vision can be noticed at altitudes of around 5000 feet when the ability to see faint sources of light will diminish.

When a climb is continued, the body is able to compensate for the lowered oxygen pressure (and therefore the amount being transferred from air to bloodstream) by increased breathing and pulse rates; 10,000 feet is generally considered the level at which these involuntary reactions are adequate. If the altitude is increased to 14,000 feet, the process reaches the limits of adequacy, and adverse effects begin to show up: Even in bright daylight, vision is impaired; thinking gets sluggish; the ability to make simple calculations and corrections deteriorates; overall performance is considerably less than optimum.

One way to get around the problems of decreasing atmospheric pressure and of allowing higher-performance airplanes to operate safely at altitudes beyond the limits of human tolerance is to *pressurize* the entire passenger compartment—that is, to create an artificial environment that makes your body think it is still at sea level, or at least no higher than 10,000 feet, where it can function normally. All jet aircraft are pressurized so that the efficiencies of very-high-altitude flight can be realized without sacrificing passenger comfort—you'd have to wear oxygen masks all the time in an unpressurized airplane.

For the airplane that is capable of flying more efficiently at altitudes beyond the tolerance of the human system and for which pressurization isn't economically justifiable, the occupants can be protected by supplemental oxygen supplied through individual masks; that is a bit uncomfortable, but it is an acceptable compromise in the light of increased speed and range and the ability to fly over most of the nasty weather.

The rules for *unpressurized* airplanes require that all occupants be provided with supplemental oxygen whenever the flight altitude exceeds 15,000 feet. But what's good enough for the occupants is not good enough for the pilot, who must use oxygen all the time above 14,000 feet, and whenever he flies for more than 30 minutes between 12,500 and 14,000 feet. Most pilots wisely stick their faces into an oxygen mask at any altitude above 10,000 feet, or plan flights at lower levels.

The symptoms of hypoxia are highly individual. Almost everyone will notice an increase in his breathing and pulse rates when the oxygen supply begins to fall short, but from there on it's very personal. Pilots who intend to fly at high altitudes should attend one of the altitude-effect courses offered at certain Air Force bases, where they can find out first-hand under controlled conditions just what to expect when they're flying too high. This is important because, even with pressurization, a failure of the environmental-control system or a leak can cause a sudden change to a hostile, low-pressure situation. Should the early-warning symptoms be ignored and the flight be continued in a hypoxic situation, the mind may push all problems aside and enter a state of euphoria—a bona fide "high" in which the pilot may know he's in trouble but just doesn't care.

Low-altitude pilots are not exempt from hypoxia, either; a number of conditions can bring your body to hypoxia-producing altitudes while the airplane is flying well below the magic 10,000 feet. The rules (and the general tolerance limits) are based on the reactions and demands of an "average" person, who is seldom encountered in the real world. Any physical or mental state less than "average" will decrease one's tolerance for lowered air pressure and cause the onset of hypoxia at lower altitudes, and in less time at any particular level in the atmosphere. Little problems on the ground can be magnified by altitude and become serious hazards when you're in command of an airplane. Perhaps the most significant reduction of altitude tolerance is that caused by smoking, since it has a direct, immediate, and lasting effect on the breathing apparatus; three cigarettes put your body at 8000 feet, and you may not need to add much actual altitude to that before hypoxia begins to show up.

If a little extra oxygen is good, more will be better—up to a point, of course, and increasing the rate and depth of your breathing will allay the onset of hypoxia in some situations. But going too far with personal supercharging (the official term is *hyperventilation*) can result in purging the blood of the small amount of carbon dioxide your system requires for normal operation. When the central nervous system senses such an upset in the blood chemistry, the body responds in the only way left: unconsciousness, which immediately restores slower, shallower breathing. You'll come to as soon as the proper balance is achieved. Hyperventilation is usually an aviator's problem in tense situations—when caught on top of a cloud deck, running out of gas, experiencing engine failure, and the like—and the only cure is a recognition of the problem and an immediate conscious effort to slow down your breathing and quiet your jangled nerves.

SPATIAL DISORIENTATION

The intricate network of communication within your body links the impulses from muscles and tendons, the pressures of sitting or standing, the information coming from the balance mechanism of the inner ear. When these inputs are placed against the background of visual perceptions, *orientation* is realized—you know the answer to "Which end's up?" But in the absence of visual clues, and with the other sources continuing their output of up-down-forward-back-right-left signals, your brain is swamped with conflicting signals that are impossible to interpret correctly. For example, a blindfolded pilot is completely unable to tell the difference between a smooth, coordinated turn and a smooth dive recovery; both maneuvers exert gravitational forces in excess of those normally felt, but without vision to correlate what he feels in the seat of his pants, he's out of luck.

A blindfold can simulate complete loss of visual clues and provide an impressive demonstration of the futility of trying to fly blind, but it doesn't hold a candle to the situation in which a pilot somehow gets himself into an *actual* condition of zero visibility. He has two alternatives: He can believe the instruments and let them replace the outside vision that's been shut off, or he can believe the sensations pouring in from the inner-ear and the deep-muscle sensors. The former results in complete control of the aircraft and an eventual return to visual conditions. The latter frequently makes headlines.

There is no immunity to spatial disorientation; even the most experienced pilots will lose all sense of direction within a half minute or so of losing visual clues. Flying (or, more correctly, *trying* to fly) by the seat of your pants is possible for a few seconds, until the airplane is displaced in any one of its three axes. Even a small movement of your head will set the fluids of the inner ear spinning, and when you exert control pressures to correct what you *think* is happening, the movements get larger, the false sensations grow in authority, and within a very short period of time the situation becomes irreversible—the airplane is out of control.

Fortunately, there are preventive measures. The pilot who is not trained in the use of flight instruments must never venture into low-visibility conditions. (It's not only illegal, it's dangerous—and especially easy to do at night, when clouds are hidden in the dark skies.) The instrument pilot must train himself to *believe* the instruments after he's learned to use them, because there will be many times when the mind would rather use information from the bodily senses than what's perceived on the instrument panel. The eyes must prevail, and the inner battle is often hard-fought.

OTHER PHYSIOLOGICAL PROBLEMS

Noise is, has been, and probably always will be an annoying dimension of flight, particularly in light aircraft, where the cost and weight of enough soundproofing to provide a comfortable environment cannot be justified. The noise generated by engines and propellers and the passage of air around the cabin at cruising speed is quite high. Pilots have accepted this many-decibeled situation as something that "comes with the job"—

and it does, but along with irreparable damage to the hearing of everyone exposed to such noise levels. The annoyance and the aural damage can be lessened somewhat with ear protectors, several styles of which are available at most airport sales counters. They range from expensive complete headsets to simple cylinders of foam, but any kind of ear protection is better than none; the consequent reduction in "nuisance" noise and improvement in radio reception as well as inside-the-cabin conversations is remarkable.

Drugs of any kind—medicinal or otherwise—are bound to impair your ability to think and act with the speed and precision required for most piloting tasks. The rules say you may not operate an airplane while using any drug that affects your faculties in any way contrary to safety, and that covers a multitude of sins. Tranquilizers, sedatives, and antihistamines can cause trouble for pilots, because of their popularity and availability. Their side-effects are predictable and manageable on the ground, but in the sky the slowed reaction times and drowsiness they often produce may be magnified enough to create genuine problems. Some patent-medicine painkillers and cough syrups—even the lowly aspirin—can have strange effects at cruising altitude.

Ear-blockage problems are essentially no different in an airplane from what they are in the office-building elevator in which you have to swallow a couple of times on the way down, but the difference in air pressure between the surface and 7000 or 8000 feet or more is a lot greater than from the fiftieth floor to the lobby. The pressures on either side of your eardrums are normally equalized by venting through the Eustachian tubes at the back of the throat. On climb-out, air flows easily from the inner ear, and, if anything, you'll notice an apparent clearing of the ears; on the way down, the higher-pressure air returns to the inside of the eardrum through the same tubes, but swallowing or yawning may be necessary to flex the openings and allow free passage of air. When the throat is inflamed, air may be blocked entirely, and the resulting pressure difference inside the head produces discomfort, probable pain, and possibly a rupture of the eardrum as the air pressure is forced to equalize. A sore throat or a cold is a good reason not to fly.

There are no across-the-counter cures for ear blockages, so the swallow-and-yawn method is universally used. (Did you think airline stewardesses used to pass out the spearmint because the passengers had bad breath?) Today's pressurized jets, whose cabins are seldom allowed to go above 8000 feet, and whose pressure can be raised and lowered very slowly, have saved the airlines a lot of money in chewing gum.

By any of its names—airsickness, seasickness, carsickness, trainsickness—the inner discomfort caused by unusual or unaccustomed motion is just as miserable. The problem inevitably solves itself; in an airplane the solution can be messy, unpleasant, and embarrassing if the one who gets sick is the pilot. The bumping, up-and-down motion produced by most aircraft flying through turbulent air is usually the culprit, although abrupt and sustained changes of aircraft attitude can also do the job. For most people, it's simply a matter of overcoming the apprehension and anxiety of suddenly finding themselves in a position completely foreign to the normal scheme of things; time, a cool, well-ventilated cabin, and attention directed to things other than one's momentary miseries will help alleviate the problem.

EMERGENCIES IN THE AIR

Before we go any further, there's a question that must be answered: Why should a discussion of airborne emergencies be located in a chapter on human factors? It's simple enough: *Most of the accidents in aviation today are caused by pilots.* In the official records, you'll find that roughly 84 percent of the aviation mishaps in the United States are directly attributable to the pilots—they fly into weather conditions they can't cope with, they fly until all the gasoline on board is used up, they allow themselves to become hopelessly lost, and then they contrive a thousand other ways to get into trouble.

In fairness, there are isolated situations when a misfortune occurs no matter how careful or knowledgeable the aviator—that's why there's an act-of-God clause in your insurance policy. But most airborne emergencies start with the human in charge, so it seems only reasonable to attack the emergency situation problem right at the source: the mental attitude of the pilot. There are precious few emergencies in the air that can't be handled and seen through to a happy ending if the pilot involved is (1) not totally surprised by the sudden turn of events, (2) prepared mentally to take charge and make the best of a potentially bad situation, and (3) sufficiently well versed in the proper technique and procedure for the situation at hand.

What follows is a series of tips to provide a foundation for further study of ways to *stay* out of trouble, and ways to *get* out of trouble with minimum risk when an aerial adventure turns sour. There are no rules for what to do in an emergency, just advice, because there are an infinite number of pilot blunders that can lead to serious problems, and each has its own best solution. The pilot who is conscious of his personal limitations, who knows what procedures to follow when his luck runs out, is a smart pilot—he's helped himself to a generous serving of aeronautical longevity. When you reach that happy plateau (and there's always another mountain to climb), you've done what you can to optimize the most important human factor of all: *your* reaction when something goes wrong. Pilot, know thyself. *Semper paratus.*

There's one thing very certain about a complete engine failure in a single-engine airplane: The silence will be overwhelming, and you'll know right away that something you didn't want to happen *has* happened. Why the engine stopped is suddenly immaterial; you're now the pilot of a fairly efficient glider, and you'd best know how to make the most of your new source of thrust—gravity—before it's depleted.

Every flying machine has a best glide speed, which is almost always coincident with the best-rate-of-climb speed; after all, the wing couldn't care less where the thrust that provides airflow comes from. With sufficient power, the airplane will climb most efficiently at this airspeed; when it is glided at the same airspeed, the altitude loss will be at the minimum, giving you more time to make the all-important decision: Where am I going to land this newly minted sailplane? Be aware that the best-rate-of-glide speed varies with the weight of the airplane; you can glide the same distance with a heavy airplane as with a lightly loaded one, but to get those results you'll need at least a rough idea of the proper airspeed to fly. The numbers (in very light airplanes just one number, which will suffice for the small range of weights possible) are in the pilot's operating handbook. Don't wait until you're an unwilling glider pilot to find out what the best airspeed is, and don't wait until you've lost a lot of altitude—set up the airplane in its optimum glide configuration just as soon as you realize that you have an engine problem. The few feet you save while slowing down may be the few feet you need to get over that last tree or power line at the end of your glide.

Once committed to a forced landing, bring your attention to bear on the matter of selecting a suitable field. Of prime importance is a continuing awareness of wind direction on the surface. If, for example, it was out of the west when you took off, it will likely still be out of the west when the engine quits, unless you've flown far enough to be under the influence of a different weather system. In any event, there are signs of wind direction and velocity all over the countryside: smoke, flags, wind patterns on water and in fields of long-stemmed crops. Your primary consideration should be to land into the wind if at all possible, reducing groundspeed and minimizing damage to yourself and the airplane.

The smoothest, firmest field you can find reasonably aligned with the wind is obviously the best choice when there's no airport around (don't forget to look *under* the airplane—it would be embarrassing to glide into somebody's corn field when there was a paved airport directly under you), but sometimes you'll be flying over terrain that hasn't a field of any kind in sight. In this situation, pick out the smoothest spot you can find (even if it's the top of a stand of trees), fly as slowly as possible, and hope for the best.

In general, when a forced landing is imminent, it's more important to get the airplane on the ground *under control* with as little forward speed as possible; let the airplane tear itself apart if necessary, but touch down with wings level. A lot of energy is absorbed when wheels and wings get torn off, and that's energy that you and your passengers won't have to absorb; if the airplane noses over or cartwheels, the chances of surviving the crash drop remarkably. *Fly it all the way in.*

Landing on a road or highway looks like a good deal at first blush, but remember the wires, culverts, fences, and such that are part and parcel of all highways. The use of a road for an emergency landing site is advisable only if it's a better choice than any of the fields nearby.

If you have the time, let someone know that you're making an unplanned descent to the earth's surface, but if you don't have the time, it's much more important to fly the airplane than to talk on the radio. A distress call will speed help your way, but the helpers might as well stay home if you lose control and really roll the airplane up in a ball while trying to tell someone of your plight.

When you know that you're going to land somewhere other than at an airport, make sure all seat belts are pulled up as tight as they will go; unlatch the door just before touchdown so that it won't jam shut; activate the emergency locator transmitter (ELT) before impact so you're sure it will be working; if there are any loose objects lying about the cabin, get rid of them or stuff thém under a seat (they usually become unguided missiles when the airplane stops suddenly). At night, your problems multiply a bit when the engine stops. You've little choice but to head for the darkest spot you can find and hope it's an open field. Save the landing lights for the last thousand feet, and steer around the big, solid obstacles that may show up as you get close to the ground. And if you don't like what you see, just turn the light off.

BETTER, BUT NOT TWICE AS GOOD

All the mental preparation, all the good procedures and techniques of engine failure apply equally to the flyer who's got two engines pulling him through the sky. Should both the power plants quit at the same time (there are astronomical odds against it), the choice has been made—you're going to land very soon whether you want to or not.

The more likely situation is very similar to that of a twin-screw boat which has suffered the loss of one of its engines— you'll get home all right, but at a much-reduced speed. In an airborne twin, there is also some concern about whether the necessary altitude can be maintained, since the loss of one engine represents a power loss considerably greater than 50 percent of the original total. Once again, there's a best-rate-of-descent speed (the wing responds to the speed of the air passing over it, not the number of engines running), which will provide the most efficient lift production.

The twin-engine pilot must demonstrate his proficiency in the procedure and technique required to fly his airplane when one of the power plants is shut down, and the twin-engine pilot who doesn't practice frequently is fooling only himself. There *are* control and performance problems that go along with nearly all of today's twins, especially when they're heavily loaded. In general, there are fewer emergency situations involving multi-engine aircraft, but consequences of the ones that do occur are usually more serious. As before, the important human factors are mental attitude, being prepared when it *does* happen, and executing the proper procedures after one of the engines gives up the ghost.

A GOOD PILOT
NEVER GETS LOST . . . BALONEY!

There are two kinds of pilots: Those who have been lost, and those who will be. There's not a veteran aviator worth his wings who wouldn't admit under pressure that at one time or another he let his navigation deteriorate to the point where he honestly didn't know where he was. Getting lost is not a true emergency in and of itself, but the pilot who is unprepared, who thinks it can never happen and therefore has no idea what to do, frequently flies himself from a relatively harmless disorientation into more serious trouble.

If the problem is simply one of a temporary loss of navigational references, it's probably best to continue along your planned route, holding the heading that your preflight computations provided; unless the wind has changed remarkably (highly unusual in most VFR conditions), you should come across something recognizable. When you reach the estimated time for crossing a checkpoint and it's not where it should be, look diligently along an arc to right and left; it's easy to miss a landmark a couple of miles off course. If your groundspeed has changed considerably and you fail to notice a checkpoint because of the time difference, keep on going—the next one is bound to show up sooner or later, and on the way you'll likely reorient yourself by other landmarks or by radio aids.

When you really get lost—don't have any idea where you are except somewhere up in the air—the very first thing to do is *settle down*. And, while you're at it, you might as well *slow* down too; it makes little sense to charge around the sky burning up the fuel that you may need to fly your way out of this mess.

For better radio communication and navigation-signal reception, gain whatever altitude you can without flying into clouds or venturing into the hypoxia trap. While you're climbing, start hollering for help; to get the most out of each transmission, use the emergency frequency—121.5—and if anyone hears you, you'll get an answer right away. When a voice link is established, pour out your problem to whoever will listen; the more information you communicate at the outset the better. When a responsible party offers suggestions or instructions to get you out of trouble, welcome them as a drowning man grabs a lifeline—and don't let go of that communications lifeline until you are absolutely sure your disorientation has been resolved.

PLANNING IS THE KEY

Every aircraft accident is the end link in a chain of events. Without exception, you could trace back through all the circumstances leading up to an emergency or a mishap and somewhere along the line discover a factor that, if handled differently, would have resulted in a completely normal and routine flight. It's the human factor again: good judgment, knowledge of one's limitations, and a healthy regard for the environment in which a pilot operates. "I was caught by the weather" and "I thought there was enough gas in the tanks" are but two of a multitude of excuses that just won't wash. When you become a pilot, you accept complete responsibility for whatever happens to your airplane and the people in it. The best procedure, then, is to break the chain of events leading to an emergency just as soon as you see an ominous situation developing. It takes a lot of intestinal fortitude sometimes, but a delay or cancellation has got to be a better deal than the other alternative.

Planning also encompasses a pilot's preparation for the emergency that couldn't happen . . . but did. Short of the airplane coming completely unglued in the air, there aren't many emergency situations that can't be reduced somewhat by rational thinking and cool reaction. The only way to make sure that *you* will do the right things at the right time is to study, memorize, drill, practice, and think about it—until an emergency becomes just another flight procedure.

Index

Acceleration error, 142, 143–44
Acceleration pump, 94
Accelerometers, 248
Access roads, 272
Accidents, 322–23
Advection fog, 194
Adverse yaw, 34
Aerial navigation, 219–48; area, 246–47; charts for, 226–28; electronic, 236–48; elements of, 220–26; inertial, 248; and navigational computers, 228–35; and pilotage, 220–27; by radar, 244–46; and VHF direction-finding systems, 246
Aerobatics, 12
Aerodynamic braking, 136
Aerodynamic stall, 22–25
Aerodynamics, 7–37; equilibrium in, 8; and flight controls, 8, 30–37; and force vectors, 8–25; and stability, 8, 26–29
Aeronautical charts, 226–28
Ailerons, 33–34, 36; in crosswind landings, 131–32; in crosswind

takeoffs, 125–26, 127, 129; and instrument turning, 173–74
Air: density of, 40–43, 87–88, 177; heating and cooling of, 176–78; moisture in, 43, 178–79, 181; neutral stability of, 177; stability of, 177–78, 183–85
Air density, 40–43, 177; and fuel-air mixture, 87–88
Air-mass weather, 179–81, 204–6
Air miles, 123
Air pressure, 40–43; in aviation reports, 213, 214; differential relationships of, 152–53; in frontal systems, 199; global patterns of, 187–90
Air-pressure instruments, 145–56; airspeed indicator, 145–46, 151–56; altimeter, 145–51; vertical-speed indicator, 156
Air temperature, 40–43; in aviation reports, 213, 214; and carburetion, 89–90; and engine cooling, 80–81; and engine-

Automatic fuel-control unit, 103
Automatic pilots, 162–65
Automatic Terminal Information Service (ATIS), 224, 281–82
Aviation reports, hourly, 212–14
Avionics, 236

Balloons, 265, 295, 296
Banking: ailerons, 33–34; and compass turning error, 142–43; and force vectors, 20–22. *See also* Turns
Base leg, 118, 287
Batteries, 83–84
Battery cart, 96
Beacons: marker, 244; rotating, 274–75
Bernoulli's lift, 14, 15
Best-angle-of-climb, 49; on airspeed indicator, 156
Best glide speed, 324
Best-rate-of-climb, 48–51, 324; on airspeed indicator, 155–56
Biennial Flight Review (BFR), 300
Blimps, 265, 294, 296
Braking, 136
Bypass engines, 103–4

Carburetion, 84–90; improvements in, 90–92
Carburetor, 85; automatic, 90; icing of, 89–90, 91; pressure, 90; and starting procedure, 92–94
Ceilings, 212, 258
Center of gravity (CG), 62–68; aft, 65–66; forward, 62–65; and moment, 68–69; of a tail-dragger, 304; and weight balance, 61–66
Center of lift (CL), 61, 62

Centrifugal force, 12; and turns, 20–22; and weight, 12, 56
Certification and ratings, of pilots, 294–300
Checkride, 296; as BFR, 300
Chord line, 23
Cirrus clouds, 183
Clear ice, 210
Clearance, 253, 257
Clearance delivery, 282
Climb performance, 17–18, 48–51; and force vectors, 18–19; and instrument flying, 168–69, 172; and takeoff, 115
"Closed traffic," 117–18
Clouds, 181–86; cirrus, 183; coverage of, 212, 214; cumulus, 178, 183–84, 185, 186, 187; stratus, 178, 183, 184–85, 186
Cold front, 198; characteristics of, 201–2; speed of, 200
Commercial pilot, 294, 296–98
Communications: and radio failure, 269–70, 283–84; while lost, 328; and traffic control, 279–84
Compass, magnetic, 139–45, 158–60
Compass-correction card, 141
Condensation, 181–85
Coning, 9
Constant-speed props, 98–99, 101, 115
Control stick, 30, 32
Control surfaces, 30, 35
Control towers, 280
Control wheel, 33
Control Zones, 258
Cooling: of air, 176–78; of engine, 78–81
Coriolis force, 188–89
Course, in navigation, 220–22
Course-deviation indicator (CDI), 240–41, 243, 247

Cowl flaps, 80
Crab angle, 128, 130, 225, 234
Crossed controls, 127, 131
Crosswind leg, 118, 286
Crosswinds, 224, 225; landings in, 130–32; and navigational calculations, 231–35; takeoffs in, 124–30
Cumulus clouds, 178, 183–84, 185, 186, 187
Currency requirements, 300–1
Cylinder, 76, 79

Datum, 68
"Dead reckoning," 220
Delta-wings, 36
Density altitude, 42n
Descent performance, 17–18, 50; and force vectors, 22; and instrument flying, 168–69, 172
Deviation error, 140
Dew point, 182, 213, 214
Dihedral effect, 29
Direction-finding systems, VHF, 246
Directional gyro (DG), 157, 158–60, 169, 172
Directional signals, 237
Directional stability, 26–28
Distance, in navigation, 220, 222–23; calculation of, 228–30
Distance-measuring equipment (DME), 242–43, 246–47
Downslope wind, 191–92
Downwind leg, 118, 286–87
Drag, 8, 10–11; and ailerons, 34; as brake, 121; induced, 11; parasite, 11
Drift, from crosswinds, 125–28, 131–32
Drizzle, 186

Drugs, 321
Dynamic wind pressure, 152–53

Ear-blockages, 321–22
Ear protectors, 321
Effective weight, 12–13, 20, 56
Electronic navigation, 164, 220, 236–48; Instrument Landing System, 243–44; low-frequency, 236–38; radar, 244–46; very-high-frequency, 236, 238–43. *See also* Pilotage
Elevators, 32–33, 36; in landing, 121; in takeoff, 113–14; and weight and balance, 61–66
Elevon, 36
Emergencies, 323–29; engine failure, 121–22, 323–27; lost, 327–28
Emergency frequency, 328
Emergency locator transmitter, (ELT), 326
Empty weight, 58–59, 68
Engines, 74–75; bypass, 103–4; and carburetion, 84–92; cooling of, 78–81; failure of, 99–100, 121–22, 323–27; fuel-injected, 90, 94–96; ignition, 81–84; jet, 102–4; lubrication of, 77–79; priming of, 93–94; principle of, 75–77; reciprocating, 74, 75–77; starting procedure for, 92–96; turboprop, 104
EPR gauge, 103
Evaporation, 178
Exceeding the critical angle of attack, 23–25

Fan engines, 103–4

Federal Aviation Administration (FAA), 67, 262, 267, 299–300
Federal Aviation Regulations (FARs), 262–70; flight safety, 264–68; general operation, 263–64; and pilots, 293–94, 300–1; and VFR, 268–70
Final approach leg, 118, 287
Fires, and over-priming, 93
Fixed-pitch prop, 97, 101, 115
Fixed stabilizers, 30
Flaps, 37; and drag, 11; and landing, 118–19; and lift, 16; and speed, 153–54
Flare, 120
Flight, force vectors in, 16–22
Flight controls: and aerodynamics, 8, 14, 30–37; primary, 30–35; secondary, 35
Flight instructors, 294, 295, 299–300
Flight instruments, 138–65; air-pressure, 145–56; airspeed indicator, 145–46, 151–56; altimeter, 145–51; attitude indicator, 157–58; automatic pilot, 162–65; directional gyro, 157, 158–60, 169, 172; and flying blind, 166–74; gyroscopic, 157–62; heading indicator, 157, 158–60, 169, 172; magnetic compass, 139–45, 158–60; turn indicator, 160–62; vertical-speed indicator, 156
Float planes, 275
Flush rivets, 11
Flying tail, 32
Fog, 185, 193–95
Force vectors, 8–25; drag, 8, 10–11; in flight, 16–22; lift, 8, 13–17; thrust, 8, 9–10; and turns, 20–22; weight, 8, 11–13
Formation flying, 265
Freezing rain, 209–10

Frisé ailerons, 34
Fronts, 195–204; characteristics of, 199–201; cold, 198, 200, 201–2; development of, 197–98; occluded, 198, 203; polar 195–96; stationary, 198–99, 200, 203–4; warm, 198, 200, 202–3
Frost, 210
Fuel-air mixture. See Carburetion
Fuel-consumption calculations, 230
Fuel-injection, 90, 94–96
Full-feathering prop, 99, 100

G force, 56–57
G loading, 56
Gasoline, 75
Generator, 84
Glide-slope signal, 119–20, 243–44, 273
Glider rating, 294, 295, 296
Gliders, 265
Governor system, for props, 98–99
Gravity, 12–13, 56–57
Ground controller, 280
Ground effect, 55–56, 114
Ground fog, 193–94
Ground miles, 123
Groundspeed, 224; calculations of, 228–38
Gusts, 211
Gyroplanes, 294
Gyroscope, 157
Gyroscopic instruments, 157–62; attitude indicator, 157–58; heading indicator, 157, 158–60, 169, 172; turn indicator, 157, 160–62
Gyroscopic precession, 157, 160

Lift, 8, 13–17, 28; Bernoulli's, 14, 15; and flight controls, 30–37, reaction, 14–15; and stalling, 22–25; and turns, 20–22
Liftoff, 114; in crosswinds, 129–30. *See also* Takeoffs
Light-gun signals, 270, 283–84
Lighter-than-air rating, 294
Lighting, at airports, 274–78
Lines of flux, 142
Load factor, 13, 56–57
Loading, 54–72; for balance, 61–67; computation of, 67–72; max gross, 57–59; maximum zero-full weight, 60 and *n*; and planning, 51–52, 59; useful load, 58–59; and weight limits, 54–61. *See also* Weight
Local controller, 280
Localizer, 243, 273
Longitude, 221
Longitudinal stability, 28
Low-Altitude Enroute Charts, 227
Low Altitude Training Routes, 259–60
Low-frequency radio navigation, 236–38
Low pressure systems, 189–90
Low-visibility flying, 166–74; climb, 168–69; descent, 168–69; turns, 173–74
Lubrication of engine, 77–79

Mag switches, 82
Magnetic compass, 139–40; acceleration error in, 142, 143–44; deviation error in, 140; and directional gyro, 158–60; oscillation error in, 140; turning error in, 142–43
Magnetic dip, 142
Magnetic north pole, 221–22

Magnetic variation, 221–22
Magnetos, 82
Manifold-pressure gauge, 98–99
Marker beacons, 244
Maximum allowable gross weight ("max gross"), 57–59
Maximum demonstrated crosswind component, 129
Maximum structural cruising speed, 155
Maximum zero-full weight, 60 and *n*
Medical examiners, 295
Military operations area (MOA), 259–60
Millibars, 213
Mixture control, 87; and engine-starting, 94–96
Moment, 68–69
Mountain wind, 191–92
MSL (mean sea level), 147

National Weather Service, 211
Nationwide planning chart, 226–27
Nautical miles, 222–23
Navigation. *See* Aerial navigation; Electronic navigation; Pilotage
Navigational computers, 228–35; and performance altitude calculation, 42–43, 230–31
Navigational satellites, 248
Negative G, 13
Neutral stability, 177
Newton, Isaac, 9, 12, 15
Noise, 320–21
Non-directional beacon (NDB), 236–38, 273
North, true and magnetic, 221–22
Nose: in crosswind landing, 130–32; in crosswind takeoff, 125–

Stability: and aerodynamics, 8, 26–29; directional, 26–28; longitudinal, 28; roll (banking), 28–29
Stall, 22–25, 117; and aft-CG, 65, 66; and airspeed indicator, 153–54; and forward-CG, 61–63; and prop angle of attack, 96
Standard altitude, 42n
Standard-day altitude, 42
STARs (Standard Terminal Arrival Routes), 228
Starters, 83
Stationary front, 198–99; characteristics of, 203–4; speed of, 200
Statute miles, 222–23
Stratus clouds, 178, 183, 184–85, 186
Structural icing, 210
Student pilot, 294, 295
Superchargers, 91
Swinging the compass, 141

Tachometer, 97
Tail-draggers, 101, 302–14; landings in, 303, 304, 308–13; takeoffs in, 303–8, 313; wheel landings in, 311–13
Tailwinds, 45–46, 224, 225; and navigational calculations, 231–35
Takeoff leg, 286
Takeoffs: and aft-CG, 65–66; crosswind, 124–30; directional control on, 111–13; force vectors in, 18; and forward-CG, 64–65; light-gun signals for, 284; maximum-performance, 132–35; normal, 110–16; short-field, 133–35; in a tail-dragger,

303–8, 313; and traffic control, 280, 281–82, 287–88; and wind, 45–47, 116, 124–30
Taxiways, 272, 273; lighting of, 276
Terminal Control Area (TCA), 256–57, 258, 282–83
Terrain modification, of air masses, 179–81
Throttle, 85–87; and engine-starting, 93–96; and manifold-pressure gauge, 98; and short-field landings, 135–36
Thrust, 8, 9–10; asymmetric, 101–2, 169; and prop action, 96–101; reverse, 100; in turbojets, 103–4
Thunderstorms, 206–9; stages of, 207
Time, in navigation, 220, 224–26; calculations of, 228–30
To-From indicator, 240–41
Tornadic tubes, 208–9
Tornadoes, 193
Torque, 102, 169
Trade winds, 188
Traffic patterns: for landing, 117–18; right-of-way in, 265–66
Transition Area, 258
Transponders, 245–46, 257
Trim tabs, 35; and instrument flying, 168–69
Trough, of low pressure, 189, 199
Turbojets, 103–4
Turboprops, 104
Turbosuperchargers, 91–92
Turn control, in autopilot, 164
Turn indicator, 157, 160–62
Turning error, 142–43
Turns: and force vectors, 20–22; and instrument flying, 173–74; level, 18. *See also* Banking
Type ratings, 294